Universitext

G. Buntebarth

Geothermics

An Introduction

With 66 Figures

Springer-Verlag
Berlin Heidelberg New York Tokyo 1984

Dr. GÜNTER BUNTEBARTH, Institut für Geophysik,
Techn. Universität Clausthal Postfach 230, D-3392 Clausthal-Zellerfeld

Translators:
INGA M. CHAPMAN, 1471 Gilmer Dr., Salt Lake City, Utah 84105,
Prof. Dr. DAVID S. CHAPMAN, Department of Geology & Geophysics,
University of Utah, Salt Lake City, Utah 84112

Title of the German Edition:
GÜNTER BUNTEBARTH, Geothermie
© by Springer-Verlag Berlin Heidelberg 1980

ISBN-13: 978-3-540-12751-2   e-ISBN-13: 978-3-642-69323-6
DOI: 10.1007/978-3-642-69323-6

Library of Congress Cataloging in Publication Data. Buntebarth, G. (Günter), 1942 –. Geothermics. Translation of: Geothermie. Includes bibliographical references and index. 1. Earth temperature. 2. Geothermal engineering. I. Title.
QE509.B9613 1984   551.1'4   84-1344

Media Conversion: Daten- und Lichtsatz-Service, Würzburg

Preface

The constantly growing demand for energy, as well as the realization during the past decade that fossil energy reserves to satisfy ever-increasing energy consumption are limited, have helped, as part of the search for alternative energy sources, to bring the subject of geothermics to its present level of significance.

Practical geothermics is concerned with prospecting for and development of geothermal heat. General geothermics deals with the thermal state of our Earth as a whole. Both divisions of this field, however, contribute practical insights, and improved methods of temperature estimation have helped to give us a better picture of detailed thermal conditions.

It is difficult for readers interested in this field to obtain an overview from the numerous, specialized papers that have been written on geothermics. This book is meant to provide a thorough introduction to the subject, although the coverage is not exhaustive is detail.

Geothermics is taught at universities and technical institutes, as part of the curriculum in geology. This introduction to geothermics is directed especially to students of geophysics and is meant to be used as a supplement to their lectures.

Special thanks for the completion of this work must be given to my teacher, Prof. Dr. O. ROSENBACH. His lectures in geophysics inspired my interest in geothermics, which is still my main research area.

Clausthal-Zellerfeld   G. BUNTEBARTH

Contents

Introduction

In ancient times it was believed that in the center of the earth there was a fire, which would break through the crust here and there and would send a visible messanger in the form of lave from the depths of the earth. This belief in a central fire was continued throughout the middle ages. Not until Descartes [6] was the development of the earth considered from the standpoint of mechanics. He compared the earth with the stars and tried to deduce from this the earth's development, giving light to the present condition of the earth's interior. Leibniz [18] thought that the center of the earth was a hot molten mass and the crust was formed during the cooling process. Newton [21] considered the shape of the earth, namely a flattened spheroid rotating about its axis, to be a direct proof of the development of the earth from a hot molten mass. Following Newton's time there were many attempts to link the history of the earth with surface features.

About the end of the eighteenth century, Buffon [4] in a rather ingenious way enlivened the discussions of the origin of the earth. From his astronomical observations of the planets' paths he deduced that the planets were once part of the molten sun and had formed after a collision of a comet with the sun. It was then that the independent existence of the earth began. In order to simulate the thermal evolution of the earth, Buffon experimented with cooling red hot iron spheres and calculated the first thermal history of the earth. In the comparison he extrapolated that the earth would remain red hot for 3000 years and that the present comfortable temperatures would be reached after 74,800 years. According to Buffon the cooling off process continues and the earth would have reached the freezing temperature of water after 93,000 years.

Fourier's work on the theory of heat formed the presently accepted foundation of thermal studies and threw new light on the thermal conditions of the earth's interior. The possibility of drawing up corrections for daily and annual temperatures of the earth's surface, plus the development of the thermometer in the first third of the nineteenth century, yielded many temperature estimates for the near-surface regions of the earth.

It is indeed evident from today's point of view that with the assumption of a central fire in the inner earth and a cooler outer surface, there is a temperature increase with depth. This is a physical necessity. Even so, many scholars disagreed until around the mid-seventeenth century. Through the observations of miners it was determined that there was a general heat increase with increase in depth. But 150 years later, when the first important book on the „Physics of the Earth" was published, the theoretical physicist Parrot [22] objected, because with increased depth in the oceans one observes a temperature decrease rather than increase. This fact was soon refuted

2

[16] and used as a proof of a general temperature increase, because the temperature at the bottom of the oceans might correspond everywhere to the maximum density of the water.

The first half of the nineteenth century was the time of change in geosciences. The advances in physics also contributed to the development of the thermal studies of the earth. Numerous temperature observations in boreholes, springs and mines, plus the extensive travels of scholars throughout the world contributed to the treatment and understanding of the earth as a whole, and individuals sought to fit their own observations into common synthesis. Particularly meritorious was Alexander von Humboldt, his observations and interpretations putting a permanent stamp on geosciences. To honour Humboldt, in 1837 Bischof dedicated his first monograph on thermal science of the earth to him with the tribute „dem umsichtigen Begründer und unermüdlichen Beförderer unserer Kenntnisse von den Temperatur-Verhältnissen der Erde“ – i.e. to the founder and untiring promoter of our knowledge of the thermal state of the earth [2].

At this time, specialized areas, which are today part of Geophysics, were still considered as a part of Physics. The term „Geophysics“ was introduced by Fröbel in 1834 and geophysics became an independent discipline about 50 years later. Geothermics was considered essentially part of general thermal studies in the first half of the 19th century. Undoubtedly, the growing number of individual observations in nature as well as experimental investigations helped to bring about a stronger differentiation in scientific subjects such as Physics, Geology and Geography. New specialities and with that a new vocabulary were formed. An obvious necessity for differentiating the thermal conditions on the earth surface developed. It had been common practice to quote the air temperatures for surface temperatures and to give their distributions at the earth surface in form of isotherms [3, 11]. In 1829, and with the use of numerous data, Kupffer compiled a map of earth temperatures and air temperatures, and determined that the two quantities do not generally coincide. On his map he represented the lines of similar earth temperatures as isogeotherms [17]. With that the vocubulary of geothermics was born. Naumann introduced the specific word „geothermics“ in his 1849 textbook on geognosy. Although he used the word to mean the temperature of the earth's interior, in his textbook he emphasized the thermal state of the uppermost crust.

The new thermal considerations and calculations were used by Hopkins [10] in his study of precession and nutation of the earth's body. He increased the size of the earth's crust, which until then had been believed to be a thin layer, to $1/4 - 1/5$ of the earth's radius. According to his concept, the solidification of the liquid mass was brought about by a cooling process from the surface as well as by increasing pressure from the interior, including the central point of the earth.

Aepinus [1] favored the development of a cold earth, formed from an accretion of meteorites. Only after the formation, according to his view, did the planet begin warming up due to an accumulation of solar energy. This hypothesis was refuted by Cassini and de la Hire [5, 8] upon the discovery of the neutral layer in the basement of the Parisian observatory. Both proved that solar energy only penetrates into the earth a few tens of meters as an annual temperature wave. The increase of temperature with depth, according to de la Rive [23] and Lyell [20] as well as Hunt [24], must be brought about chemically. Already at the turn of this century the contribution of radioactive elements to the temperatures of the interior of the earth were being discussed. Extensive analysis proved that radiogenic heat production is present in all rock forms [25]. These data, together with hypotheses concerning the chemistry and structure of the earth in particular assuming an exponential decrease of radiogenic heat production with depth, enabled the calculation of temperature distributions within the earth [9, 13, 14]. Estimations of the temperature in the earth's interior with values between 2000 and 10,000 °C were already made at the beginning of this century. In comparison, an actual core temperature between 4000 and 5000 °C is most likely according to present day knowledge of the properties of core materials and the extrapolated results of laboratory experiments. This will be shown in Chapter 4.

Today much more detailed information is available on the temperature distributions of the crust and upper mantle. All the disciplines of geosciences contribute to this knowledge. Considerable credit is due to the paradigm of new global tectonics, particularly stemming from the contributions of Wegener. The present picture of the thermal conditions of the earth's interior has also been influenced strongly by the results of the research on rocks, minerals and metals at elevated temperatures and pressures. Changes in crystal structures and the behavior of the physical parameters give important clues about the condition and structure of the earth's interior. The temperature dependence of the rates of reactions can be also used as an indicator of a thermal state. Thus changes in physical properties can often be used to deduce temperatures. Chapter 5 introduces frequently applied methods for estimating temperatures from the interior of the earth.

Besides general geothermics, the applied aspects of this field have been important in this century, as will be shown in the last chapter. Methods of exploring for heat reservoirs were developed. The usefulness of geothermal heat is still in it's initial phase but in the near future it will gain greater importance.

1 Physical Basis of Heat Transfer

Temperature is one of the most important physical properties of the earth. It is both spacially variable and time-dependent, on a variety of time scales, from the variation of surface temperature in the course of a year to the thermal evolution of the entire planet, which takes billions of years.

The different temperatures are the result of both lateral and vertical temperature differences on a small scale as well as on the scale of the entire earth. These temperature differences will be equilibrated through heat transport. But this equilibration can only take place at a finite speed, so that this process is time-dependent. On a large scale this is so slow that it cannot take place undisturbed. The slow but steady movement of the plates, in a plate tectonic sense, and the formation of mountains, rifts and throughs with their accompanying magmatism influence the temperature distribution of the earth. Heat sources in the earth's interior are continually being distributed and created by mechanical forces. Temperature differences can be decreased, increased for equilibrated through this forced mass transport.

1.1 Temperature and Temperature Gradient

At a point specified by the position vector $\vec{x}$ and at a given time t the temperature can be expressed as

$$T = f(\vec{x}, t). \tag{1.1}$$

One is free to choose any stationary coordinate system in space and time. The temperature scale is equally independent. A set of points so specified with their temperatures constitutes a temperature field which, when built from scalar quantities, is itself a scalar field. If the points of this field take on the same value for the function T, namely

$$T = \text{constant}$$

they form an equipotential surface, or in this case an isothermal surface. If the temperature field is reduced to a two dimensional representation, the lines with $T = \text{const.}$ are called isotherms. If the isothermal surface T_1 intersects the point $\vec{x}_1$ and at a distance $|\Delta\vec{x}|$ the isothermal surface T_2 intersects $\vec{x}_2$, then one can express the temperature increase from $\vec{x}_1$ to $\vec{x}_2$ as

$$\frac{\Delta T}{\Delta\vec{x}} = \frac{T_2 - T_1}{\Delta\vec{x}} \quad \text{if } T_2 > T_1. \tag{1.2}$$

The limit for an infinitesimal interval at point $\mathring{x}_1$ is called the temperature gradient at $\mathring{x}_1$:

$$\operatorname{grad} T = \lim_{\Delta \mathring{x} \to 0} \frac{f(\mathring{x}_1 + \Delta \mathring{x}, t) - f(\mathring{x}_1, t)}{\Delta \mathring{x}} . \tag{1.3}$$

The gradient of the temperature field is a vector quantity, which is defined at every point of the field and points normal to the isothermal surface in the direction of increasing temperature.

Together in space with the temperature field T, the temperature gradients constitute a gradient field

$$\operatorname{grad} T = g(\mathring{x}, t) \tag{1.4}$$

which like temperature is dependent both on position and time. The dimensions of the geothermal gradient are generally given as [°C/km]. On a global scale the temperature distribution of the earth consists of a field in which the isothermal surfaces are represented by spherical surfaces. The earth's surface is the isothermal surface with the minimum temperature and the center of the earth the point of maximum temperature.

1.2 Heat Flow Density, Thermal Conductivity and Thermal Diffusivity

If at any given point $\mathring{x}_1$ in space there is a temperature gradient different from zero, then an equilibrating process occurs which contributes to the diminishing of the gradient, provided no additional heat sources or sinks are present at point $\mathring{x}_1$. During the equilibration process heat flow is transported following in the direction of the temperature gradient. This energy flow, normalized in respect to time and area, is called heat flow density $\vec{Q}$.

$$\vec{Q} = - \mathbf{K} \operatorname{grad} T. \tag{1.5}$$

Heat flow density is a vector quantity. Thus in space, heat flow density is a vector field, just as the temperature gradient is a vector field.

The magnitude of heat flow density is proportional to the temperature gradient, whereby the proportionality factor is defined as the thermal conductivity ($\mathbf{K}$). Thermal conductivity is a property of the material in which the heat is transported. In general, the thermal conductivity in crystal materials is a tensor quantity ($\mathbf{K}$), however, in crystals of the cubic crystal symmetry such as garnet, rocksalt and galena it can be reduced to a scalar quantity, so that only the components K_{11}, K_{22} and K_{33} of the tensor $\mathbf{K}$ are different from zero and have the same value. A body with this property is called isotropic. However, most rock-forming minerals such as quartz, feldspar, and mica are anisotropic. With a statistical collection of anisotropic crystals through polycrystalization the body as a whole behaves as an isotropic medium (see Section 2.1).

6

Besides the thermal conductivity K there is an associated parameter called thermal diffusivity which is identified by the symbol κ and defined as the quotient of thermal conductivity K and the product of density ϱ and specific heat c.

$$\kappa = \frac{K}{\varrho c}. \tag{1.6}$$

The thermal diffusivity κ has the dimension (m^2/s).

1.3 The Heat Conduction Equations

In a given body one can consider an infinitely small cylinder, which is bounded by two flat surfaces dF perpendicular to the cylinder axis and separated by the distance dn. The cylinder encloses the volume dV. Inside this cylinder there is a homogenous isotropic heat source A, which generates within volume dV, in unit time dt the following quantity of heat $d^2 q^*$:

$$d^2 q^* = A \, dV \, dt. \tag{1.7}$$

Part of the heat quantity $(d^2 q_1^*)$ is that which increases the heat content in dV by increasing the temperature

$$d^2 q_1^* = \varrho c \, dV \, dT. \tag{1.8}$$

The difference $d^2 q_2^* = d^2 q^* - d^2 q_1^*$ corresponds to the quantity of heat

$$d^2 q_2^* = -\left(K \frac{\partial T}{\partial n} \right) dF \, dt \tag{1.9}$$

which flows through surface dF in a unit time, namely:

$$A \, dV \, dt = \varrho c \, dV \, dT - \left(K \frac{\partial T}{\partial n} \right) dF \, dt. \tag{1.10}$$

With the help of the Gauss' theorem one converts the surface integral to a volume integral and obtains:

$$A \, dV \, dt = \varrho c \, dV \, dT - \operatorname{div}(K \operatorname{grad} T) \, dV \, dt \tag{1.11}$$

and equation (1.11) leads to the differential equation of heat conduction:

$$\varrho c \frac{\partial T}{\partial t} = \operatorname{div}(K \operatorname{grad} T) + A. \tag{1.12}$$

Applying differential operators

$$\Delta T = \operatorname{div}(\operatorname{grad} T) \quad \text{Laplace-Operator}$$
$$\nabla T = \operatorname{grad} T \quad \text{Hamilton-Operator}$$
$$\quad (= \text{Nabla-Operator})$$

the heat conduction equation takes on the following form:

$$\varrho c \frac{\partial T}{\partial t} = \nabla K \nabla T + K \Delta T + A. \tag{1.13}$$

The heat conduction equation describes the temperature field in space and time in an isotropic medium with a spatially dependent, but temperature-independent thermal conductivity.

In the case of constant thermal conductivity the equation is reduced to

$$\varrho c \frac{\partial T}{\partial t} = K \, \varDelta T + A \tag{1.14}$$

and with the use of thermal diffusivity

$$\frac{\partial T}{\partial t} = \kappa \, \varDelta T + \frac{A}{\varrho c}. \tag{1.15}$$

In a steady state condition i.e. constant in time $\partial T/\partial t = 0$, one obtains Poisson's equation,

$$\kappa \, \varDelta T + A/\varrho c = 0 \tag{1.16}$$

which in the case of negligible heat production ($A = 0$) further reduces to Laplace's equation

$$\varDelta T = 0. \tag{1.17}$$

Analytic solutions to the heat conduction equation are possible only in simple cases. These cases are determined by initial and boundary conditions. The initial condition specifies the temperature distribution at time zero corresponding to the beginning of the mathematical formulation. Note that the zero point in time does not have to coincide with the beginning of a physical process, for example with the intrusion of magma.

The most frequent initial condition ($t = 0$) in geothermic problems is

$$T = \text{constant} \quad \text{for all } \mathring{x},$$

e.g. the temperature of the wall rock during an intrusion.

The boundary conditions are conditions in space, which are valid generally at the edges or surfaces of the model. The boundary conditions may be time-dependent.

Frequent boundary conditions in the treatment of geothermic problems are:

1) constant surface temperature on the models ($T_0 = $ constant for $t \geq 0$), e.g. the mean annual temperature of the earth's surface,
2) periodic temperature changes on the surface ($T = T_0 \sin \omega t$ for $t \geq 0$), e.g. diurnal temperature changes on the earth surface,
3) constant heat flow density ($Q = $ constant for $t \geq 0$), e.g. the heat flow density from the upper mantle which is considered constant within one tectonic province.

Poisson's and Laplace's equations are relatively easily integrated for one-dimensional problems. However, the one-dimensional analytical solution limits the choice of models because:

1) Only parallel layered, homogenous, isotropic rocks can be considered.
2) The thermal diffusivity must be constant within each layer.
3) The heat source distribution must be described by an analytic function within each layer.

8

The main heat sources for geothermic problems are radiogenic heat producers (uranium, thorium and potassium). These can be neglected ($A = 0$), can be constant within a layer ($A = A_0$), or can be a function of position, e.g. $A = A_0 \exp(-x/H)$.

The integration of Poisson's equation with the condition $A = A_0 \exp(-x/H)$ yields;

$$T(x) = T_0 + x Q_0/K - x \frac{A_0 H}{K} \exp\left(-\frac{h}{H}\right) + \frac{A_0 H^2}{K}\left(1 - \exp\left(-\frac{x}{H}\right)\right) \qquad (1.18)$$

where:

T_0 is the temperature of the top surface,
Q_0 is the heat flow density through this top surface,
h is the thickness of the layer,
A_0 is the heat production at the upper surface,
K is the thermal conductivity and,
H is the particular distance at which heat production is reduced to the value $1/e = 0.368$ of its value at the top boundary.

If the heat production $A = A_0$ is constant, equation (1.18) reduces to:

$$T(x) = T_0 + \frac{1}{K} Q_0 x - \frac{A_0}{2K} x^2. \qquad (1.19)$$

Finally if $A = 0$ we have the solution of Laplace's equation:

$$T(x) = T_0 + \frac{1}{K} Q_0 x = T_0 + x \operatorname{grad} T \qquad (1.20)$$

in which the temperature gradient is constant.

Supplementary Problems

1.1 Determine an analytical expression for heat transfer using the general heat conduction equation for steady state (Eq. 1.16) and substituting a linearly decreasing heat generation rate with depth ($A = A_0(1 - z/2H)$).

1.2 Estimate the heat flow density through the Mohorovičić discontinuity, at a depth of $z = 30\,\text{km}$, if an exponential law of heat generation rate is applied ($A = A_0 \exp(-z/H)$), and if a linear law is applied $A = A_0(1 - z/2H)$. What difference is calculated, if $H = 7.5\,\text{km}$ and $A = 4.2 \cdot 10^{-6}\,\text{W/m}^3$?

1.3 Determine the temperature difference at the Moho for steady state using Eq. 1.18 assuming an exponentially decreasing heat generation rate and using the solution to problem 1.1, where $A_0 = 4.2\,\mu\text{W/m}^3$, $K = 3\,\text{W/m}\,^\circ\text{K}$, $H = 7.5\,\text{km}$, and $z = 30\,\text{km}$.

1.4 Repeat 1.3 using Eq. 1.18 and Eq. 1.19 for a constant heat generation rate $A_{00} = A_0 H/z$.

2 Thermal Properties of Common Rocks

2.1 Thermal Conductivity

Thermal conductivity K is an important physical quantity in the transport of heat. It both controls temperature gradients in individual layers of the earth's crust under stationary conditions and determines the time scale for transient processes such as the cooling of intrusive bodies.

Thermal conductivity is defined for the stationary condition of conductive heat transfer as the quotient of heat flow density, i.e. energy flow per unit area, and the temperature gradient in a one-dimensional heat conductor.

$$K = \frac{Q}{dT/dx}.$$ (2.1)

Values of thermal conductivity of rocks under conditions found at the earth's surface vary between about 1 and 6 W/m °K with a few notable exceptions. Table 2.1 gives values for a selection of rocks exhibiting this variation in thermal properties. The defined scalar quantity K is a material property, dependent not only on the type of rock or mineral but on the crystal structure as well which might cause an anisotropy in thermal conductivity. The anisotropy itself causes a dissipation of heat with different rates in different directions, and the direction of heat flow does not have to coincide at any point with the highest temperature gradient. Anisotropy arises not only from the arrangement of ions in a crystal structure, but also in a macroscopic scale in rocks exhibiting a preferred orientation of individual mineral grains. Rocks

Table 2.1. Thermal conductivity (K) and thermal diffusivity (κ) for as number of materials under normal conditions [2.15, 2.17]

Material	K [W/m °K]	κ [10^{-6} m^2/s]
Limestone	2,2–2,8	1,1
Slate	2,4	1,2
Sandstone	3,2	1,6
Bituminous coal	0,26	0,15
Rock salt	5,5	3,1
Gneiss	2,7	1,2
Granite	2,6	1,4
Gabbro	2,1	–
Peridotite	3,8	–

with a distinct texture, such as sedimentary rocks and many metamorphic rocks, display a definite anisotropic behavior.

Finally between the earth's surface and the earth's interior there are great temperature and pressure differences so that thermal conductivity cannot be considered as a constant. Its dependence on temperature as well as on pressure must be taken into account.

The thermal conductivity of a rock can be estimated from the conductivities of the constituative minerals and the fraction of each mineral phase. A maximum conductivity is calculated from the weighted arithmetic mean value $K_{max} = \sum_i p_i K_i$, and a minimum value from the weighted harmonical mean value $1/K_{min} = \sum_i p_i/K_i$ where p_i is the fraction of the i-th mineral having the conductivity K_i and $\sum_i p_i = 1$.

Since quartz is a good conductor of heat, the rocks bearing this mineral exhibit a strong dependence of the thermal conductivity on the fraction of quartz, which in the case of granite, vary from about 2.5 to 4 W/m °K according to a fraction of 20 to 35% quartz [2.2]. An increase of the plagioclase proportion, especially of anorthite-rich plagioclase lowers the thermal conductivity of a rock due to the low heat conductivity of this mineral [2.2, 2.12].

2.1.1 Temperature Influence on Thermal Conductivity

One starts with the assumption that thermal conductivity in the interior of the earth is governed by two mechanisms: the lattice or phonon conductivity by radiation. A part of each component is dependent on the temperature. From room temperature to several hundred degrees the propagation of thermal energy, which is due to unharmonic lattice interactions, dominates in rocks. This means that in electric insulators such as crystals the phonon conductivity dominates.

This conductivity K_L should be inversely proportional to absolute temperature T

$$K_L \sim \frac{1}{T}. \tag{2.2}$$

Experiments [e.g. 2.16, 2.25], have confirmed this relationship and have shown that thermal conductivity of rocks up to about T = 700 °C can be well expressed by the function:

$$1/K_L = a + bT \tag{2.3}$$

where a and b are constants.

In the simplest layered models one can separate the continental lithosphere, i.e. the region of the earth from the surface down to a maximum of two one hundred kilometers, into three layers. These are the silica rich upper crust, the intermediate to basic lower crust and olivine rich upper mantle. With this global generalization one can describe average conductivity functions as follows [2.1, 2.25]:

$$
\begin{aligned}
\text{upper crust:} &\quad K_L^{-1}[\text{m °K/W}] = 0.33 + 0.33 \cdot 10^{-3} \, T[°C] \\
\text{lower crust:} &\quad K_L^{-1}[\text{m °K/W}] = 0.41 + 0.29 \cdot 10^{-3} \, T[°C] \\
\text{uppermost mantle:} &\quad K_L^{-1}[\text{m °K/W}] = 0.21 + 0.5 \;\cdot 10^{-3} \, T[°C].
\end{aligned} \tag{2.4}
$$

This last equation describes only the phonon contribution, but temperatures in the upper mantle are sufficiently high that the contribution to conductivity through radiation must also be considered.

This radiative contribution K_R for an olivine-rich upper mantle is given by [2.25].

$$K_R[W/m\,°K] = -0.52 + 2.3 \cdot 10^{-3}\,T[°C] \quad \text{for } T \geqq 230\,°C. \tag{2.5}$$

The total conductivity results from the sum of the two contributions

$$K = K_L + K_R. \tag{2.6}$$

For single crystals of olivine, one generally measures higher values [e.g. 2.9] than are calculated from equation (2.6). These values can be considered as preferred thermal conductivities for coarse crystalline material or assumed to be present in the lower part of the upper mantle, rather than the lower values obtained with fine crystalline rocks.

Experimentally determined radiative conductivity increases linearly with temperature. Consequently it always remains far behind the theoretically expected temperature influence, because theory predicts that radiative conductivity should rise according to the third power of temperature. The deviation can be explained in part through scattering at grain boundaries of minerals and through radiation absorption by iron atoms in the infrared regions.

Because of the experimental results, thermal conductivity models can be made for the earth's interior down to about 400 km depth [for example 2.25]. Conductivity at that depth amounts to about double the olivine conductivity at room temperature (Fig. 2.1). Slight modifications have to be made for differing models of the continents and oceans. Crustal thickness is about 30 km under the continents but less than 10 km in oceanic regions, thus the thermal conductivity differences are due to material contrasts as well as temperature differences.

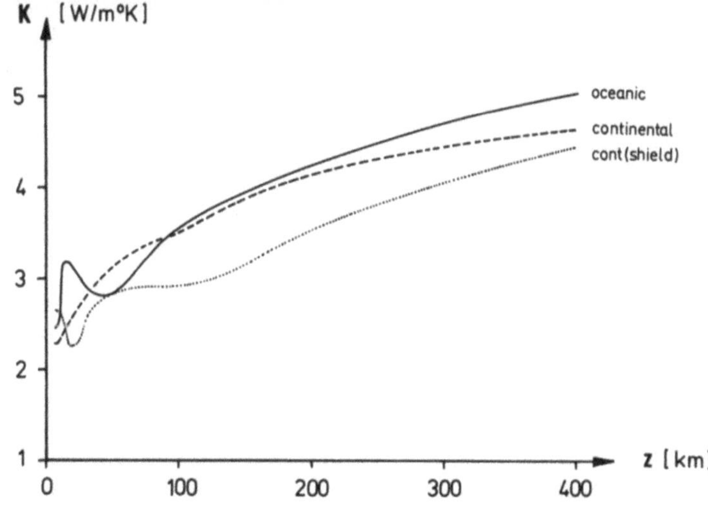

Fig. 2.1. Model of thermal conductivity of the earth's interior for continental regions (normal type, old shields) and oceanic areas

12

Within this general lithosphere thermal conductivity model, a special case must be made for quartz which is a widely occurring mineral in rocks of the upper crust. Because of its common occurrence and its physical property of changing symmetry from trigonal alpha-quartz to hexagonal symmetric beta-quartz, this mineral has to be considered separately. The temperature of symmetry transformation is dependent on pressure p [MPa] and is given by

$$T[°C] = 573 + 0.26\,p. \tag{2.7}$$

Such temperature can be reached in certain regions of the continental crust at depths where granite or gneiss (about 30% quartz) occurs. The phase transformation expresses itself in the change of physical properties, such as seismic velocity and thermal conductivity. The transition temperature in quartz bearing rocks is reported to increase with $0.6-0.7\,°C/MPa$ confining pressure instead of $0.26\,°C/MPa$ for single crystals of quartz [2.30]. The changes in thermal conductivity can amount to a decrease of over 20% [2.16] and result in a temperature increase below the depth in which the phase transformation takes place.

Apart from the change in crystal symmetry, quartz and feldspar can be mobilized in the presence of water under certain temperature-pressure conditions – a fact which also leads to the reduction of thermal conductivity.

2.1.2 Pressure Influence on Thermal Conductivity

Under low pressure all rocks possess a porosity consisting of pore spaces between individual mineral grains and microcracks which occur both between and within grains. With increasing pressure porosity gradually decreases and above about 100 MPa its influence is small. Even though the porosity of granite, gneiss and mica schist is only on the order of 1%, their physical properties such as sound velocity and thermal conductivity are considerably altered with the closure of pores. The pressure correction for crustal rocks (granite, gabbro, gneiss etc.) at pressures up to 100 MPa is typically about 10% [2.13, 2.32].

Under greater pressures the elastic properties of individual crystals, through deformation of the crystal lattice, influence the thermal conductivity. Thermal conductivity increases with increasing compression. This increase is linear with pressure p up to the elastic limit:

$$K = K_0(1 + a\,p) \tag{2.8}$$

where a is on the order of magnitude $(1 \text{ to } 5) \cdot 10^{-5}\,\text{MPa}^{-1}\ (= 1 \text{ to } 5\,\text{Mbar}^{-1})$ [2.2, 2.26].

More recent measurements [2.28, 2.29] of thermal diffusivity κ show a similar behavior, for the entire pressure region of $0-300$ MPa and can be represented by equation (2.8):

$$\kappa = \kappa_0(1 + â\,p) \tag{2.9}$$

where â takes on values from $(1 \text{ to } 5) \cdot 10^{-4}\,\text{MPa}^{-1}\ (= 10 \text{ to } 50\,\text{Mbar}^{-1})$ for crustal rocks.

2.1.3 Thermal Conductivity of Anisotropic Bodies

With minerals and rocks having a directionally preferential thermal conductivity, the heat flow density is described by the following equation

$$\vec{Q} = - \mathbf{K} \operatorname{grad} T. \tag{2.10}$$

Instead of the scalar quantity K the tensor $\mathbf{K}$ is used, which possesses three independent components of thermal conductivity in the direction of the three perpendicular coordinate directions x, y and z. The conductivity components can be measured either on single crystals or on rocks with distinct structure, one perpendicular (K_z) and one parallel (K_x) to the layering. In the last case one assumes $K_y = K_x$. The random arrangement of anisotropic mineral grains for strongly deformed gneiss etc., results in a mean value which can be estimated, hence producing a mean scalar value for the quantity K.

The various methods give respectively the maximum and minimum average values of:

$$K_{max} = 1/3 (K_x + K_y + K_z) \tag{2.11}$$

$$K_{min} = 3 (1/K_x + 1/K_y + 1/K_z)^{-1}. \tag{2.12}$$

The commonly used geometric average

$$K_g = \sqrt[3]{K_x K_y K_z}$$

lies in the region of

$$K_{min} \leqq K_g \leqq K_{max}.$$

With anisotropic minerals, the anisotropy of the crystal structure frequently expresses itself in the habit of single crystals. They are elongated or have a laminated appearance such as quartz, tourmaline, mica and other minerals. On a different scale the layering and compositional changes in sedimentary rocks results in a large anisotropy of their physical properties. An example is shale (see Table 2.2). This anisotropy is maintained during metamorphism. Rocks of distinct structure exhibit a large differ-

Table 2.2. Ratio (R) of the largest to the smallest thermal conductivity for a number of minerals and rocks at room temperature

Mineral/Rock	$R = \dfrac{K_{max}}{K_{min}}$	References
Quartz	2,1	[2.14]
Feldspar (Orthoklase)	1,1	[2.24]
Olivin ($Fo_{92} Fa_8$)	2,0	[2.18]
Orthopyroxene	1,9	[2.18]
Slate	2,5	[2.15]
Mica shist (alpine)	1,4	[2.33]
Granite (alpine)	1,1	[2.33]
Dunite	1,3	[2.18]

ences in thermal conductivity perpendicular $\perp$ and parallel $\parallel$ to the stratification [2.10].

On the other hand magmatic rocks often show little or very minor anisotropy which can generally be ignored in geothermal work.

2.1.4 Thermal Conductivity of Porous Rocks

In addition to crack porosity, rocks can also have a volume porosity which does not change with unidirectional pressure or under compression. This cannot be neglected for sedimentary rocks. Sandstones at 1–2 km depth often possess a porosity of up to 15%. Because pores at depth are filled with oil, water or gas, which in contrast to the matrix of the rock has a lower thermal conductivity, the effective conductivity of the total rock can be greatly reduced.

Depending on whether pore spaces can be taken as isolated volumes, or as an interconnected system, one obtains different estimates for the thermal conductivity of the entire system. The first model (Fig. 2.2a) yields a maximum value, and second (Fig. 2.2b) a minimum value.

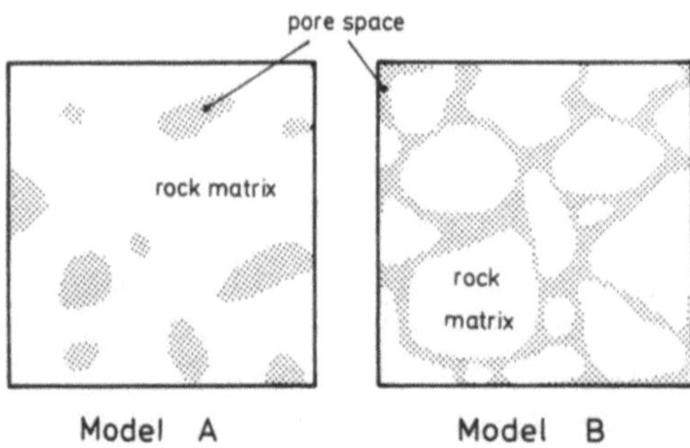

Fig. 2.2. Model of pore spaces for estimating the total thermal conductivity

For small porosities effective thermal conductivities are given by the following values [2.32]:

Maximum value (Model A):

$$K = K_M \left[1 - \frac{3\,\Phi(1 - K_P/K_M)}{2 + \Phi + K_P/K_M} \right]. \tag{2.13}$$

Minimum value (Model B):

$$K = K_M \left[1 - \frac{\Phi(1 + 2K_P/K_M)(1 - K_P/K_M)}{\Phi(1 - K_P/K_M) + 3K_P/K_M} \right]. \tag{2.14}$$

K_P is a thermal conductivity of the pore-filling fluid and K_M is that of the rock matrix.

An overview of the influence of pore-filling fluid on the conductivity of sandstone with its varying porosities is given in Table 2.3.

Table 2.3. Estimation of thermal conductivity K of sandstone of varying types of pore fluids at $T = 300\,°K$ ($K_{quartz} = 6.1\ W/m\ °K$)

Model of the porespace	K [W/m °K] with pore filling					
	Water		Oil		Gas	
	$\Phi = 5\%$	15%	$\Phi = 5\%$	15%	$\Phi = 5\%$	15%
A	5,7	5,0	5,8	5,2	5,7	4,9
B	5,2	3,9	5,7	4,9	1,2	0,4

2.2 Specific Heat

The increase of the internal energy (q*) of a volume element is proportional to its mass (m) and the temperature (see Section 1.3). The proportionality factor is called specific heat c and is given by:

$$c = \frac{1}{m}\frac{dq^*}{dT} \tag{2.15}$$

with dimension [Ws/g °K].

For rocks which are not porous, the average specific heat amounts to $c \approx 0.8\ Ws/g\ °K$, a value which has a significant dependence on the temperature. For crystalline rocks, this temperature dependence at constant pressure is given by the following equation:

$$c_p[Ws/g\ °K] = 0.75\,(1 + 6.14 \cdot 10^{-4}\,T - 1.928 \cdot 10^4/T^2) \tag{2.16}$$

where temperature T is the absolute temperature.

Sedimentary rocks often have a high porosity and when they are saturated with water, the corresponding specific heat increases because of the relatively high specific heat of water ($c = 4.2\ Ws/g\ °K$ at $T = 293\,°K = 20\,°C$. Within the upper crust the specific heat of water can reach double its value (i.e. $c = 8\ Ws/g\ °K$ at $T = 350\,°C$ and $p = 20\ MPa$).

In the case of saturated porous rocks, a specific heat can be calculated using a weighted average from the values of the matrix and the pore-filling fluid. In Table 2.4

Table 2.4. Specific heat (c) of a few materials at $T = 20\,°C$ [2.17]

Material	$c\left[\dfrac{Ws}{g\,°K}\right]$
Sandstone	0,71
Calcareous sandstone	0,84
Clay	0,86
Bituminous coal	1,26
Oil	2,1
Ice	2,1
Water	4,2

the specific heat of a few materials is given which are commonly used in geothermic problems.

Under the very high pressures and temperatures in the upper mantle and especially in the core one should consider not only the specific heat at constant pressure (c_p) but also the specific heat at constant volume (c_v). This is especially relevant for calculations of convection in the earth. Both specific heats are interrelated and their ratio is given by

$$c_p/c_v = 1 + \alpha \gamma T \tag{2.17}$$

in which α is the volume coefficient of expansion, T the absolute temperature (°K) and γ the Grueneisen-parameter which lies in the range of $1 \leq \gamma \leq 2$.

2.3 Radiogenic Heat Production

2.3.1 Radioactivity of Rocks on the Earth's Surface

Some concentration of radioactive elements can be detected in all rocks. Radiated energy from nuclear decay is converted to heat by means of absorption. The most important elements are uranium, thorium, and the unstable isotope ^{40}K in naturally occurring potassium. Rubidium (^{87}Rb) also belongs to this group of heat producers, but because of its small concentration and small energy production rate (10^{-12} W/g of the natural rubidium) contributes so little to the total heat production (about 1%) that it is mentioned only for completeness. Radiogenic heat production A is calculated from the concentrations c of U[ppm], Th[ppm] and K[%] after [2.21] as:

$$A[\mu W/m^3] = \varrho (9.52\, c_U + 2.56\, c_K + 3.48\, c_{Th}) \cdot 10^{-5} \tag{2.18}$$

where $\varrho\,[g/cm^3]$ is the density of the rock.

With the exception of potassium, the other two elements in general do not form independent minerals. Uranium and thorium minerals are of lesser significance because of their rare and local occurrence, in comparison to the large areas where uranium and thorium are found as trace elements. Only in their global distribution are they significant to the total heat of the earth. These elements occur as trace elements and potassium possibly as a mineralizer in the magmatic, metamorphic and sedimentary rocks. Even though the concentration in individual samples of matching and similar rock types can vary greatly, one can detect a general rule. The amount of uranium, thorium and potassium increases with higher silica concentration of the rock.

These three elements can be bound differently in rocks. The binding mechanisms result in two principal constituents, namely an easily soluble and a less soluble component. The less soluble part is fixed in the crystal lattice of one or more individual minerals, in inclusions or ionic exchange positions incorporated or built into the mineral structure. The soluble part is that which is absorbed at grain boundaries or crystal surfaces and/or is found in pore spaces of a rock.

The soluble constituent of radioactive elements increases proportionality with their total concentration (Fig. 2.3) and becomes mobile as soon as water migrates

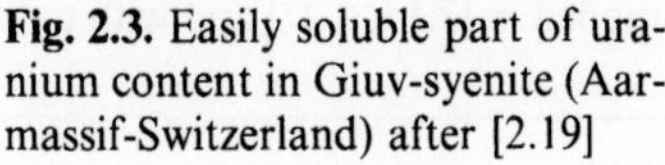

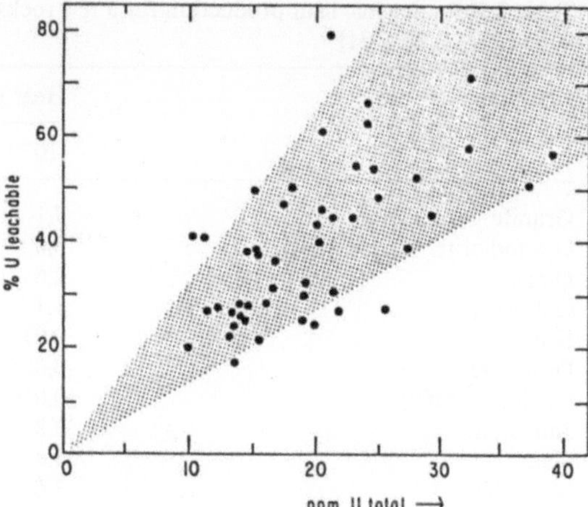

Fig. 2.3. Easily soluble part of uranium content in Giuv-syenite (Aarmassif-Switzerland) after [2.19]

through the rock. This migration leads to a redistribution especially of uranium, but also of thorium. Such redistribution takes place not only during weathering but also during the cooling of intrusions in the upper crust [2.5, 2.23] and during rock metamorphism. This latter process may even lead to a redistribution of the relatively soluble components.

The high mobility of these lithophile elements is a reason for their increased abundance in the upper layers of the earth's crust, a process which can be observed throughout the earth's history and which has not yet been completed. The spatial and the chronological distribution of radioactive heat producers greatly influences the temperature field of the inner earth, because one must assume that $\frac{1}{3} - \frac{1}{2}$ of the surface heat flow density is produced by the decay of unstable isotopes.

The concentration of heat-producing elements in individual rock types (Table 2.5) lets one assume a heat production distribution which is determined mainly by the rock type. However, it has to be assumed that within a homogeneous lithologic layer of the earth's crust the radioactive heat production is not constant. This realization came with the discovery of rock radioactivity [2.11]. An upper crust with a constant heat production typical of an average value for granite would be sufficient to produce the entire heat flow from the earth. But since it has to be assumed that there is a contribution to the heat flow from the deeper interior, the heat production in homogeneous layers cannot be constant or at least cannot correspond to the average values of the respective rock, as they are measured at the earth's surface.

2.3.2 Methods for Estimating Radioactive Heat Production in the Earth's Interior

For the description of heat production in rocks of the magmatic differentiation sequence from gabbro to granite, the silica content alone is adequate. This dependency is insufficient, however, for rock containing metallic ores or a large component of spinel. It is also not sufficient if the same chemical composition is present at different pressure modifications as for example with gabbro and eclogite.

Table 2.5. Radiogenic heat production for a few rocks as compiled in [2.15, 2.27] and with data from [2.21, 2.22, 2.31]

Rock type	Heat generation	
	$[10^{-13}$ cal/cm^3 s]	$[\mu W/m^3]$
Granite	7,1	3,0
Granodiorite	3,6	1,5
Diorite	2,6	1,1
Gabbro	1,1	0,46
Dunite	0,01	0,0042
Peridotite	0,025	0,0105
Olivinfels (Eifel)	0,036	0,015
Sandstone	0,8–2,4	0,34–1,0
Slate	4,4	1,8
Mica shist	3,6	1,5
Gneiss	5,8	2,4
Amphibolite	0,8	0,3
Eklogite		
low U-content	0,08	0,034
high U-content	0,35	0,15
Chondrite (Stone meteorite)	0,063	0,026

One has to choose a mineral specific quantity which is independent of the chemical process. For this purpose one can conceive the density ϱ of a mineral as comprised of an anion part (ϱ^a) and cation part (ϱ^c). Furthermore, assume that the anions are only oxygen ions which lie as closely packed spheres, in which case the anion part (ϱ^a) is nearly constant. The density is influenced mostly through the cation part (ϱ^c). The compressional wave velocity (v_p) is also a function of the cation distribution in the crystal lattice of many minerals [2.6, 2.23]. This arrangement of the cations is called the cation packing index [2.5, 2.8].

From the relationships between the cation packing index (k-value), density ϱ and also the compressional velocity v_p, a good correlation can be determined between k-value and the radioactive heat production A [2.4, 2.5, 2.23]. The above mentioned dependencies:

$$\text{k-value} \circ\!\!-\!\!\circ \varrho$$
$$\text{k-value} \circ\!\!-\!\!\circ v_p$$
$$\text{k-value} \circ\!\!-\!\!\circ A$$

establish a correlation between compressional wave velocity (v_p), density (ϱ) and heat production (A):

$$\varrho \circ\!\!-\!\!\circ A$$
$$\varrho \circ\!\!-\!\!\circ v_p \quad [2.20]$$
$$v_p \circ\!\!-\!\!\circ A \quad [2.21].$$

The relationship to heat production applies especially to crystalline rocks. No data for sedimentary rocks were used.

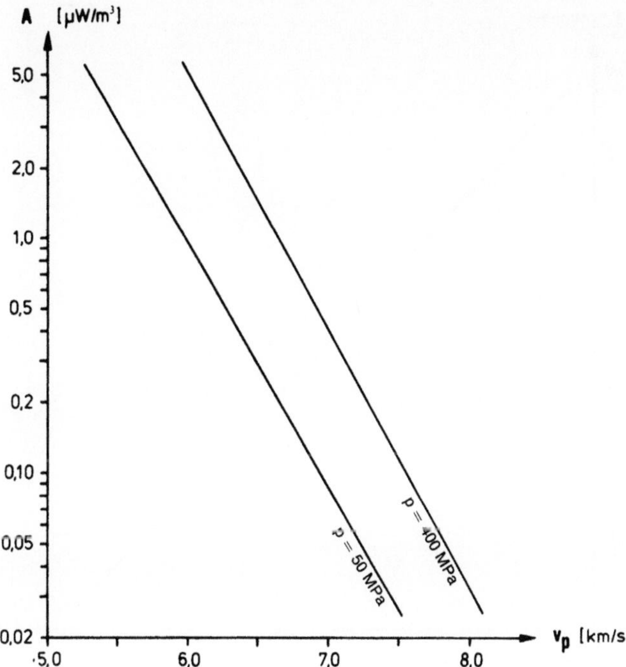

Fig. 2.4. Relationship between heat production A and sound wave velocity v_p for pressures p = 50 MPa and p = 400 MPa

With the use of the above correlations it is possible to utilize measured seismic velocities results and/or gravity model calculations in estimating the heat distribution within the earth's crust [2.5].

The seismically determined compressional wave velocity is a quantity which is not only dependent on rock type but also on the temperature and pressure at depths where the velocities have been determined. In order to set up a relationship between velocity and heat production, the pressure and temperature influences have to be considered. In Fig. 2.4 the $A - v_p$ – relationship is given for normalized conditions. In actual use, seismically obtained velocities v_p must be multiplied by a correction factor according to depth [2.4] (see Table 2.6). In the correction factor, average temperatures for continental areas are included. For zones of reduced velocity, e.g. caused by partial melting, Fig. 2.4 is not applicable.

Table 2.6. Factors for pressure corrections and temperature corrections of sound wave velocity v_p for use of Fig. 2.4 at p = 400 MPa

Seismic velocity v_p [km/s]	Correction factor for a depth z [km] of					
	5	15	20	25	30	35
6,0–6,4	1.020	1.016	1.021	1.039	–	–
6,5–7,5	1.013	1.016	1.017	1.022	1.032	1.042
> 7,5	1.019	1.016	1.015	1.020	1.022	1.022

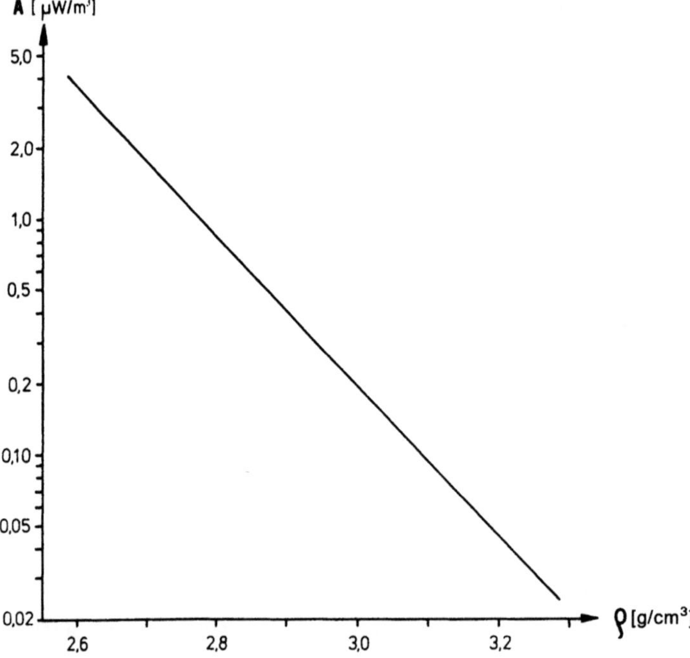

Fig. 2.5. Relationship between heat production A and density ϱ

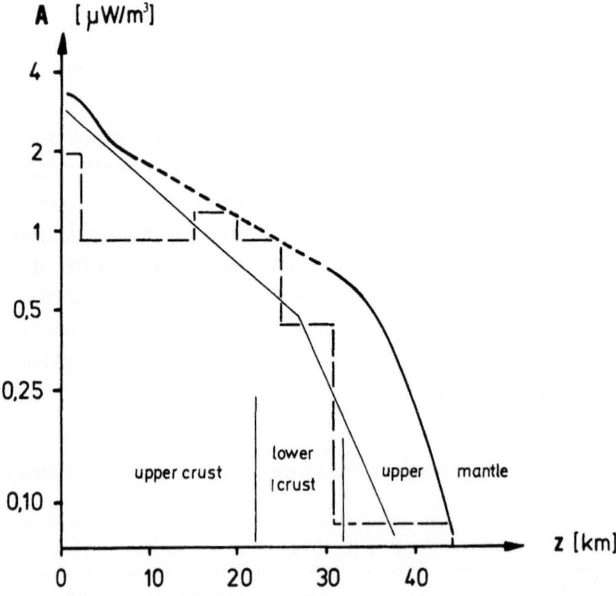

Fig. 2.6. Model for distribution of heat production in the crust of the foothills of the Alps [2.5] according to
—— seismic results
– – – gravimetric models
—— petrological model

The correlation between density and heat production (Fig. 2.5) is less pressure- and temperature-sensitive. As a result the normal conditions at the earth's surface can be transferred directly to the density within the crust.

Density-depth distributions are generally described as step functions, thus heat production distribution will also be described by such functions. On the other hand, the velocity-depth functions are often determined as monotonic functions, which on their part yield corresponding monotonic heat production distributions (Fig. 2.6). In many cases adequate gravity and seismic data are not available for the estimation of heat production. In this case a petrologic layer model has to be constructed, whereby global average values of the heat production based on rock type are assigned to the middle of each layer. Between the individual points there is a distribution of heat production which approximates an exponential function (Fig. 2.6).

The methods demonstrated above are only applicable for the crust and the uppermost part of the mantle. For greater depth there are only rough estimations available for heat production from U-, Th-, K-concentrations in ultrabasic rocks and in stony meteorites, which can be considered representative of mantle material. With a basic to ultrabasic composition for the upper mantle, assuming steady state heat flow conditions, the heat flow density at the surface would be higher than presently observed. It is suspected that only part of the heat generated in the mantle reaches the surface or else that the measured heat production is not representative of the upper mantle. If one could prove that the generated heat is greater than that which is lost from the upper surface, then there would be proof that the earth evolved from a cold state and heated up slowly.

Supplementary Problems

2.1 The pore fluid of a water-satured porous sandstone ($\Phi = 25\%$) is gradually replaced by precipitated silica from percolating thermal water. Determine the change in the heat conductivity, where the rock matrix and the silica have a conductivity of $K = 6.1$ and 0.65 W/m °K, respectively.

2.2 Estimate the difference in heat conductivity of a basalt containing 8 % magnetite and a basalt without ore. Use both weighted means: the arithmetic and the harmonic. The conductivity of basalt is $K = 2.5$ W/m °K and that of magnetite $K = 5.1$ W/m °K.

2.3 A heat generation model is given for southern Norway and Denmark on p. 55. Determine the crustal contribution of the radiogenic heat to the heat flow density at the surface, applying a linear decrease of heat generation with depth.

2.4 Repeat 2.3 for an exponentially decreasing heat generation.

3 Analytical Treatment of Conductive Cooling in the Crust

Numerous thermal problems occur in connection with tectonic and magmatic processes in the earth's crust. It is customary to describe these processes, such as magmatic intrusions or volcanic events, as thermal equilibration processes based on heat conduction. Whereas tectonic events can hardly by treated without convective heat transport, intrusions or volcanic events can be modelled with thermal equilibration based on conduction only. In geological reality this assumption is too simple. For example, in the cooling of a magma body, water migrates through the country rock and transports a considerable amount of heat. However, simplifications can and must be made in the construction of models. In doing so one must treat the cooling process with enough accuracy to produce meaningful results while taking into consideration the errors resulting from these simplifications.

For the bodies considered below there are special assumptions required which enable simple analytical solutions of the heat conduction equation to be found. The simple solution to a problem has the advantage that the model can be varied quickly and without difficulty.

Before specific models are discussed, it is appropriate to introduce a mathematical function which commonly occurs in the solution of time-dependent thermal problems:

$$f(x) = \int_0^x \exp(-a^2)\, da. \tag{3.1}$$

This function is called the error function, if defined as:

$$\mathrm{erf}(x) = 2\sqrt{\pi} \int_0^x \exp(-a^2)\, da. \tag{3.2}$$

The error function is tabulated in the appendix. It is equal to zero for $x = 0$ and one for $x \to \infty$. In practice an upper boundary of $x = 2.5$ can be used, at which point the deviation of the functional value of $\mathrm{erf}(x)$ from unity is smaller than five parts in a thousand.

With many cooling models it is assumed that the intrusion instantaneously penetrates the adjacent rock and its temperature remains constant during this process. The error which occurs with this assumption will become negligible with cooling and the passing of time.

The country rock which bounds on an intrusion or also an extrusion is assumed to be isotropic, homogeneous and to have the same thermal properties as the magma. With differences in the thermal conductivity and thermal diffusivity, heat can be transferred more slowly or quickly to the country rock. In spite of the relatively strong temperature dependence of thermal diffusivity and thermal conductivity, this dependency is generally not considered. However, this error can be minimized by using average values of the quantities over the temperature intervals being considered.

A further quantity, which should not be ignored, is the heat of fusion from magma given off during crystallization. This heat of fusion can amount to ⅓ of the heat content in a magma body. This portion decreases for hotter magmas. Part of the latent heat is, however, carried away upon the release of fluids into the adjacent rock, so that the heat of fusion really amounts to about 25% of the heat content.

A solution to the heat conduction equation is very difficult using analytical methods if one includes the heat gradually released during the crystallization process. But there are two methods for taking this amount of heat into account. First, an additional temperature can be added to the intrusion temperature. This consists of the quotient of the heat of fusion (L) and specific heat (c)

$$T^* = L/c \approx 300 \,^{\circ}C. \tag{3.3}$$

Second, the size of the intrusive body can be increased by a additional volume which would give off the same quantity of heat as the heat of fusion in the crystallizing magma.

In both cases the error decreases with passing of time. In the first instance one obtains a better approximation of the actual results for large values of time and in the second case for small values of time.

3.1 Thermal Equilibration in the Homogenous Half Space

3.1.1 Half Space with a Boundary Surface

If hydrothermal fluids or surface water migrates along a fault zone, then the temperature field in the nearby rocks is altered. If a lava flow covers the surface, the upper surface of the ground will be heated; and if magma rises or spreads laterally within the crust, the adjacent regions, will be heated. Such geological events can be modelled and the time-dependent temperature field in the country rock can be calculated with the aid of the model shown in Fig. 3.1. The basis of this model is that the temperature at the surface of the half space is constant (T_{GR}). Since, as a rule, a thermal event fades out after a certain time, T_{GR} does not remain constant in time with the cooling of

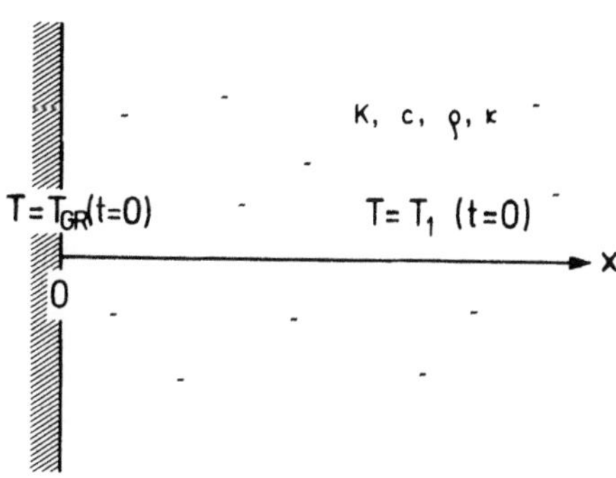

Fig. 3.1. Model of a half space with a surface boundary

24

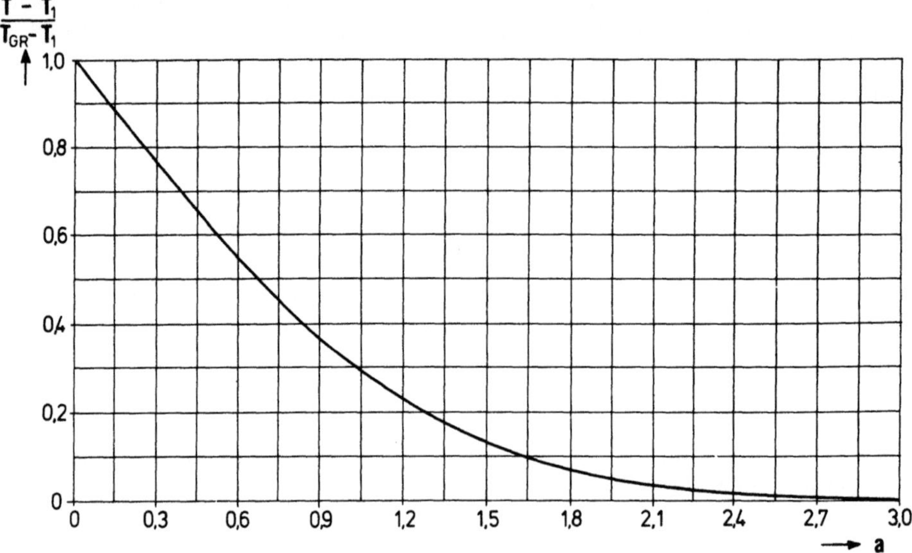

Fig. 3.2. Relative temperature in a half space with a surface boundary after equation (3.4). The dependent variable is the dimensionless parameter $a = x/\sqrt{2\kappa t}$

magma. Therefore, the model can be used only for times when T_{GR} doesn't change considerably.

If T_1 is the initial constant temperature of the country rock and T_{GR} is that which momentarily adapts itself to the boundary surface, then temperature in the country rock as a function of position and time is given by:

$$T(x, t) = (T_{GR} - T_1)\left[1 - \mathrm{erf}\left(\frac{x}{2\sqrt{\kappa t}}\right)\right] + T_1 \tag{3.4}$$

and the heat flow density Q is given by:

$$Q(x, t) = K\frac{\partial T}{\partial x} = (T_{GR} - T_1)\exp\left(-\frac{x^2}{4\kappa t}\right)\sqrt{(K c \varrho)/(\pi t)}. \tag{3.5}$$

In the country rock the total heat supplied per unit surface area amounts to:

$$\tilde{Q}(t) = \int_0^t Q(0, t^*)\,dt^* = 2(T_{GR} - T_1)\sqrt{K c \varrho t/\pi}. \tag{3.6}$$

Fig. 3.2 gives the temperature changes calculated from equation (3.4). The relative temperature $(T - T_1)/(T_{GR} - T_1)$ is plotted against the quantity $a = x/\sqrt{2\kappa t}$ which may be thought of as a dimensionless distance.

3.1.2 Subsurface with a Cover of Lava

Consider the case where lava with temperature T_1 covers part of the earth's surface, solidifies, and cools. In the process of cooling, the lava warms the uppermost layers

Fig. 3.3. Model of a half space with two boundaries (i.e. lava flow on the earth's surface)

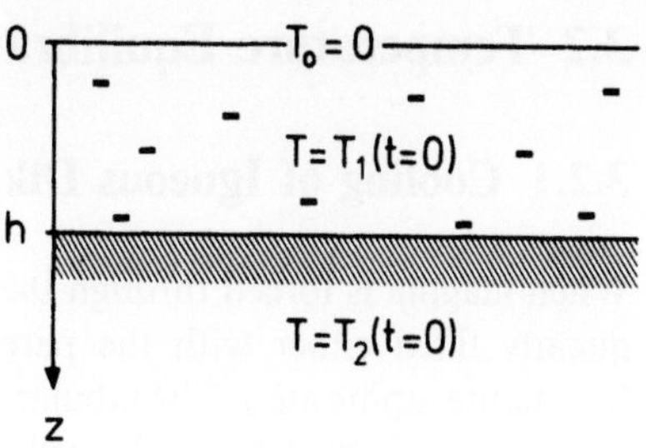

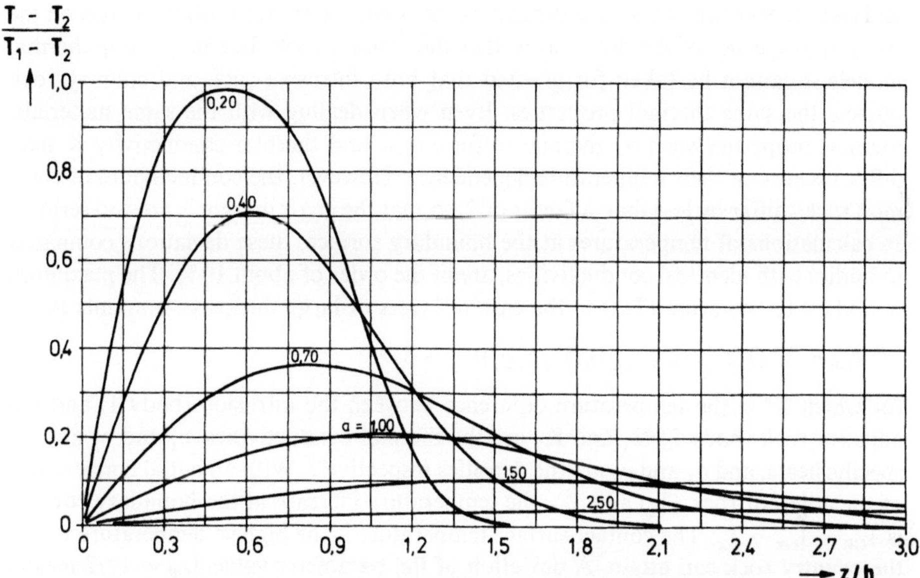

Fig. 3.4. Relative temperature in a half space with two boundaries after equation (3.7). The depth parameter is normalized by a layer thickness h. Curve parameter is dimensionless time $a = \sqrt{2\kappa t}/h$

of the subsurface having an initial constant temperature T_2. In such a case the thermal state of the region can be calculated with the model shown in Fig. 3.3. On account of the one-dimensional assumption the thickness of the layer (h) must be much smaller than its lateral extent. The lava surface (z = 0) is taken at air temperature, which for expediency is assumed to be $T_0 = 0\,°C$. The boundary surface (z = h) between the base of the lava flow and the previous earth surface should have a perfect contact so that no thermal contact resistance exists and temperature is assumed to be continuous across the contact. It is also assumed that rocks both for $0 \le z \le h$ as for $z > h$ have identical thermal properties. The temperature field is then given by equation (3.7):

$$T(z, t) = T_2 + \frac{T_1 - T_2}{2} \left[\mathrm{erf}\left(\frac{h - z}{2\sqrt{\kappa t}}\right) + 2\,\mathrm{erf}\left(\frac{z}{2\sqrt{\kappa t}}\right) - \mathrm{erf}\left(\frac{h + z}{2\sqrt{\kappa t}}\right) \right]. \qquad (3.7)$$

Fig. 3.4 shows the relative temperature change $(T(z, t) - T_2)/(T_1 - T_2)$ and its dependence on the ratio z/h. The curve parameter is $a = \sqrt{2\kappa t}/h$.

3.2 Temperature Equilibration in Model Bodies

3.2.1 Cooling of Igneous Dikes

When magma is forced through the upper crust, cracks are formed which are subsequently filled either with the parent magma itself or with one of its derivatives (pegmatite, aplite etc.). The tabular bodies, known in their solid form as dikes, cooled off and the adjacent rock is heated in the process. Depending on the temperature and the duration of this process, the adjacent rock can also be altered. The temperature and its influence are not only dependent on the distance from the dike but also on the thermal properties of the dike – as well as the country rock. In constructing thermal models it cannot be taken for granted that both intrusive rock and country rock possess the same thermal properties. Even when dealing with the same materials, thermal properties such as thermal diffusivity κ and thermal conductivity K may differ because of their temperature dependence. However, the conductivities of common rocks differ by less than a factor of 2, so that the error margin is not too serious. In calculations of temperatures at the boundary surface, these deviations compared to bodies with identical conductivities, are in the order of about 10%. The maximum boundary temperature (T_{GR}) in the case of layers of large thickness, amounts to:

$$T_{GR} = T^*(1 + \sqrt{(K_2 c_2 \varrho_2)/(K_1 c_1 \varrho_1)})^{-1} + T_2 \tag{3.8}$$

for which T^* is the temperature difference between the intrusion (body 1) and the adjacent rock (body 2) K_1 and K_2 are the thermal conductivities, c_1 and c_2 are the specific heats, and ϱ_1 and ϱ_2 are the densities respectively. With an initial temperature of the adjacent rock (T_2) and T_{GR} the temperature increase at the boundary contact is $\tilde{T}_{GR} = T_{GR} - T_2$. The contact surface temperature is the highest temperature which the country rock can attain. A deviation of the parameter value $\tilde{T}_{GR} = T^*/2$ means that the heat is conducted away faster $\tilde{T}_{GR} < T^*/2$ or more slowly $\tilde{T}_{GR} > T^*/2$ than it would, were all materials to have identical thermal properties.

Fig. 3.5 illustrates the model on which the calculations are based. The thickness (2 D) of the plate must be considerably smaller than its length and breadth in order to justify the one-dimensional consideration. If the intrusion and the adjacent rocks

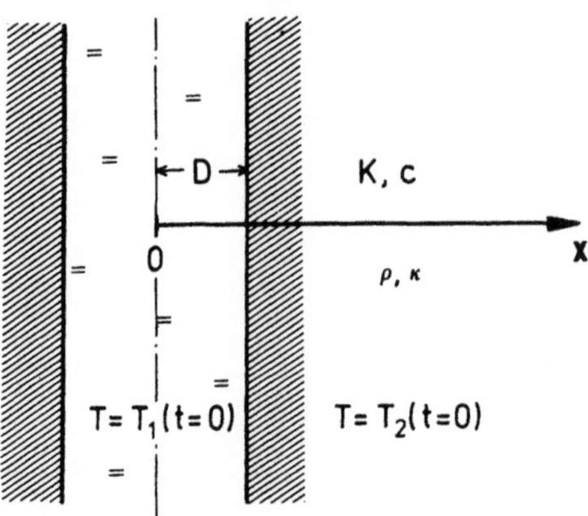

Fig. 3.5. Model of a dike

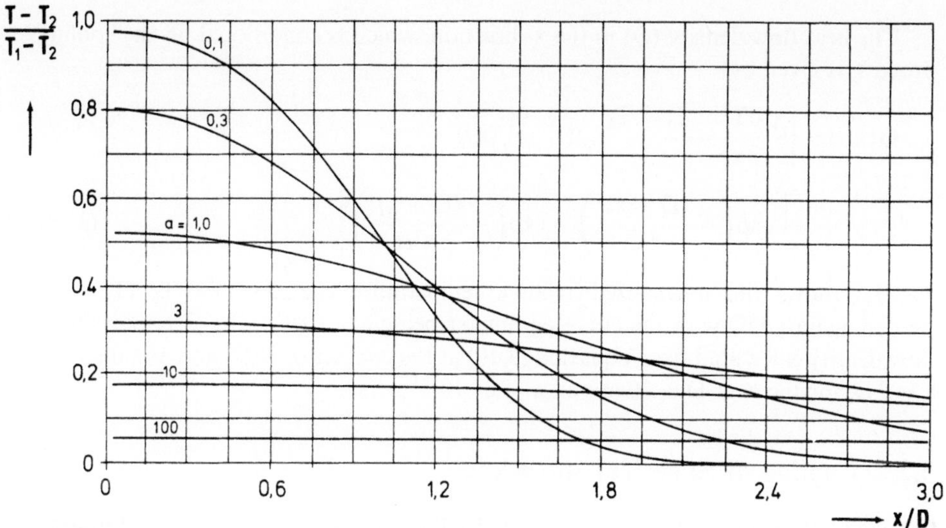

Fig. 3.6. Relative temperature in a dike and country rock after equation (3.9) and Fig. 3.5. Distance from the center of the dike is normalized by the dike half width. Curve parameter is dimensionless time $a = \kappa t / D^2$

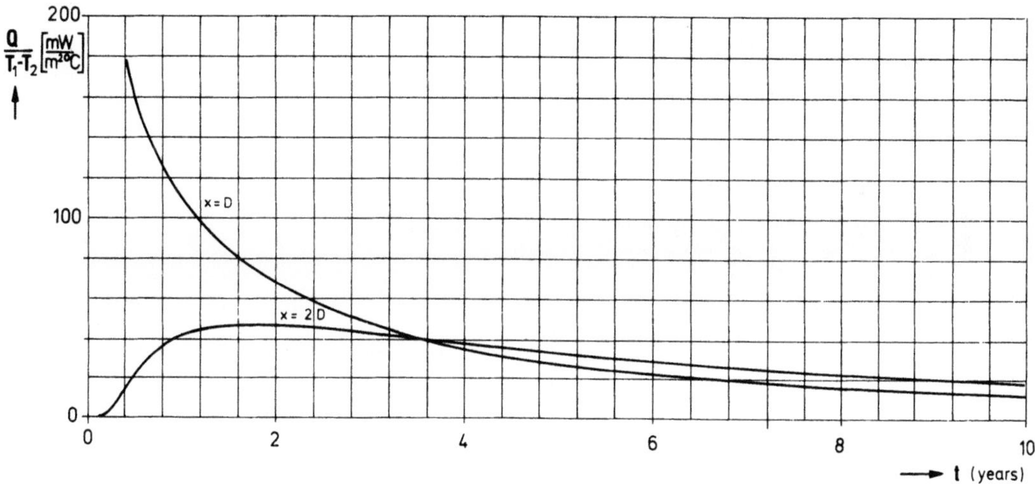

Fig. 3.7. Heat flow density away from a dike as a function of initial temperature and time after equation (3.10). The two curves represent positions at the contact ($x = D$ and at $x = 2D$). This particular model is based on a dike thickness of $2D = 20\,\mathrm{m}$ and a thermal diffusivity $\kappa = 8 \cdot 10^{-7}\,\mathrm{m^2/s}$

have the same thermal properties, then the time- and position-dependent temperature field is given by:

$$T(x,t) = T_2 + \frac{T_1 - T_2}{2}\left[\operatorname{erf}\left(\frac{D+x}{2\sqrt{\kappa t}}\right) + \operatorname{erf}\left(\frac{D-x}{2\sqrt{\kappa t}}\right)\right] \tag{3.9}$$

with the temperature of the adjacent rock (T_2) at time $t = 0$ and the intrusion temperature T_1.

28

The heat flow density (Q) in the x-direction, which is conditional on the cooling process, is given by:

$$Q(x,t) = \left| K \frac{\partial T}{\partial x} \right| = \frac{T_1 - T_2}{2} \sqrt{(K c \varrho)/(\pi t)}$$

$$\cdot \left[\exp\left(-\frac{(D + x)^2}{4 \kappa t} \right) - \exp\left(-\frac{(D - x)^2}{4 \kappa t} \right) \right]. \tag{3.10}$$

Fig. 3.6 shows the normalized relative temperature changes $(T - T_2)/(T_1 - T_2)$ plotted against x/D with the curve parameter being $a = \kappa t/D^2$. Fig. 3.7 shows heat flow density calculated from equation (3.10) at the contact $(x = D)$ and at a distance $x = 2D$, for the case $D = 10$ m and $\kappa = 8 \cdot 10^{-7} \, \text{m}^2/\text{s}$.

3.2.2 Cooling of Spherical Intrusions

Consider next a magma which is intruded into the crust, and forms a spherical shaped body. Let the intrusion have a radius R and an initial temperature T_1 at a time $t = 0$ as shown in Fig. 3.8. The corresponding initial temperature of the adjacent rock is $T = T_2$.

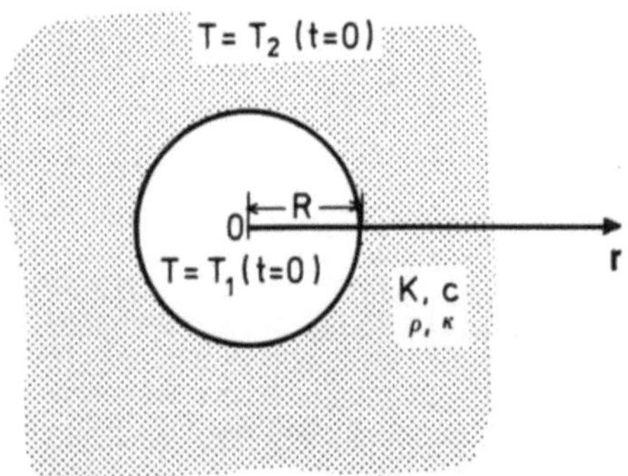

Fig. 3.8. Model of a spherical intrusion

If the intrusion and country rock have identical thermal properties, the temperature field during cooling is described by

$$T(r,t) = T_2 + \frac{T_1 - T_2}{2} \left\{ 2 \sqrt{\frac{t\kappa}{\pi r^2}} \exp\left(-\frac{(R + r)^2}{4 \kappa t} \right) - \exp\left(-\frac{(R - r)^2}{4 \kappa t} \right) \right.$$

$$\left. + \text{erf}\left(\frac{R + r}{2\sqrt{\kappa t}} \right) + \text{erf}\left(\frac{R - r}{2\sqrt{\kappa t}} \right) \right\}. \tag{3.11}$$

The corresponding heat flow density (Q) during cooling is given by

$$Q(R,t) = \left| K \frac{\partial T}{\partial r} \right|_{r=R} = \frac{K(T_1 - T_2)R}{4 \kappa t} - \frac{K(T_1 - T_2)}{R^2} \sqrt{\frac{\kappa t}{\pi}}$$

$$\cdot \left\{ \exp\left(-\frac{R^2}{\kappa t} \right) \cdot \left(1 + \frac{R^2}{\kappa t} \right) - 1 \right\}. \tag{3.12}$$

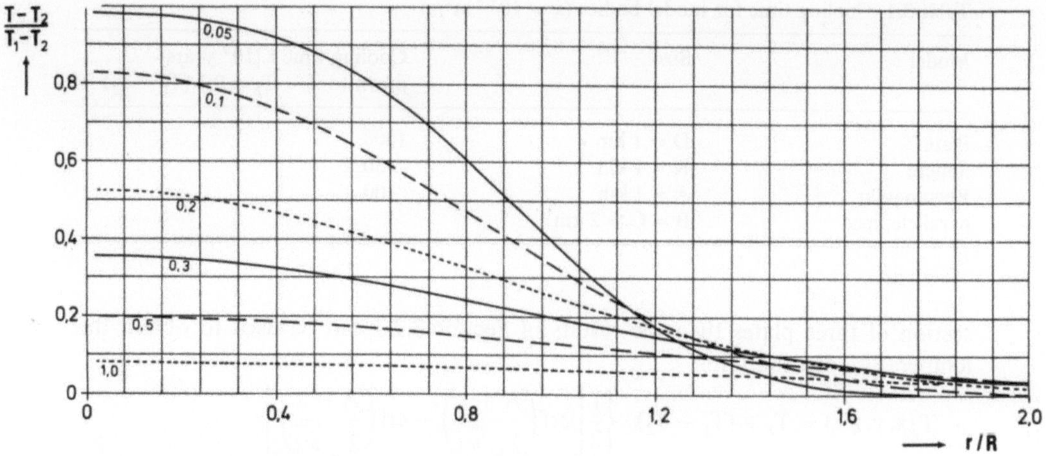

Fig. 3.9. Relative temperature for a spherical intrusion and country rock. Distance r is normalized by sphere radius R. The curve parameter is dimensionless time $\kappa t/R^2$

Both temperature and heat flow density in this idealized geometry are functions of radial distance from the center of the intrusion and of time. In Fig. 3.9 the cooling process is shown graphically in terms of the dimensionless parameter $a = \kappa t/R^2$.

3.2.3 Cooling of Rectangular Intrusions

Consider a three-dimensional rectangular shaped model as shown in Fig. 3.10. The edges have dimensions 2A, 2B and 2C in the x, y and z directions respectively.

For the purpose of computing cooling histories, many intrusions can be adequately described with such a model. If the rectangular body is considered as the inter-

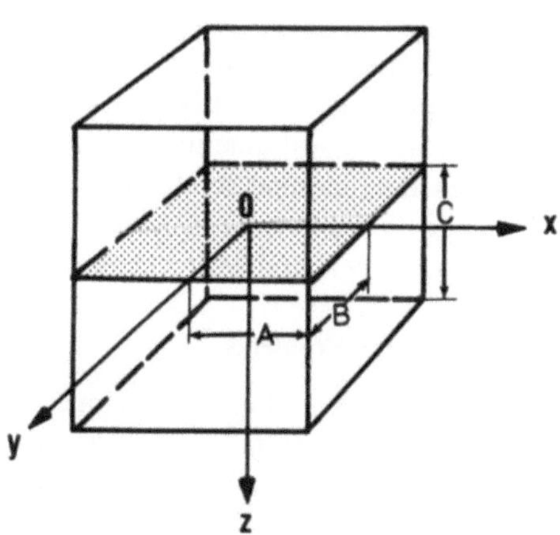

Fig. 3.10. Model of a rectangular intrusion

Table 3.1. Cooling time for model bodies ($\kappa = 10^{-6}\,\mathrm{m^2/s}$)

Model	Size	Cooling time t [10^3 years] down to $T - T_2 = 0{,}1\,(T_1 - T_2)$
Plate	$D = 1\,\mathrm{km}$	1000
Sphere	$R = 1\,\mathrm{km}$	30
Rectangular	$A = 1\,\mathrm{km}$	100
parallelepiped	$B = C = 2\,\mathrm{km}$	

section of three plates then the results of Section 3.2.1 can be used to obtain the temperature distribution

$$T(x,y,z,t) = T_2 + (T_1 - T_2) \cdot \left\{ \frac{1}{2}\left[\mathrm{erf}\left(\frac{A+x}{2\sqrt{\kappa t}}\right) + \mathrm{erf}\left(\frac{A-x}{2\sqrt{\kappa t}}\right) \right] \right.$$

$$\left. \cdot \frac{1}{2}\left[\mathrm{erf}\left(\frac{B+y}{2\sqrt{\kappa t}}\right) + \mathrm{erf}\left(\frac{B-y}{2\sqrt{\kappa t}}\right) \right] \cdot \frac{1}{2}\left[\mathrm{erf}\left(\frac{C+z}{2\sqrt{\kappa t}}\right) + \mathrm{erf}\left(\frac{C-z}{2\sqrt{\kappa t}}\right) \right] \right\}. \tag{3.13}$$

T_2 is the temperature of the adjacent rock and T_1 is the temperature of the intrusion at time $t = 0$. The heat flow density in z-direction $(Q(z,t))$ is given by

$$Q(z,t) = \left| K\frac{\partial T}{\partial z} \right| = \frac{T_1 - T_2}{8}\sqrt{(K c \varrho)/(\pi t)}\left\{ \exp\left(-\frac{(C+z)}{4\kappa t}\right) - \exp\left(-\frac{(C-z)}{4\kappa t}\right) \right\}$$

$$\cdot \left\{ \mathrm{erf}\left(\frac{A+x}{2\sqrt{\kappa t}}\right) - \mathrm{erf}\left(\frac{A-x}{2\sqrt{\kappa t}}\right) \right\} \cdot \left\{ \mathrm{erf}\left(\frac{B+y}{2\sqrt{\kappa t}}\right) + \mathrm{erf}\left(\frac{B-y}{2\sqrt{\kappa t}}\right) \right\}. \tag{3.14}$$

A graphical representation of the solution is not meaningful because of the large choice of values for the lengths A, B and C. One must perform a calculation for each special case being considered. An upper limit for cooling is given by the temperature distribution in a dike of the same thickness. For the rectangular body the temperature $T(x,y,z,t)$ is lower than that of the dike $(T(x,t))$ and the cooling time is correspondingly shorter. As an example, Table 3.1 gives the times necessary to cool the center of an intrusive body to 10% of the original temperature difference with country rock.

It should be remembered that temperatures calculated for the above models can only be considered reference points. Some quantities which strongly influence the thermal processes cannot be included or only unsatisfactorily in calculations. This includes (a) physical quantities which change in space and occasionally in time such as anisotropy, thermal conductivity, specific heat, intrusion temperature and latent heat, and (b) geological processes such as repeated magmatic pulses, collapse of magma chambers and hydrothermal processes.

Supplementary Problems

3.1 Thermal water of constant temperature has ascended a fault for 10 years. Determine the distance from the fault at which the temperature increases to 10% of the difference $(T_{GR} - T_1)$ above the temperature of the country rock. Assumed values: heat conductivity $K = 2\,\mathrm{W/m\,°K}$, specific heat $c = 1\,\mathrm{Ws/g\,°K}$, density $\varrho = 2.5\,\mathrm{g/cm^3}$.

3.2 The surface has been covered by a lava flow of thickness h = 20 m. Determine the time at which the maximum temperature is reached in a depth of z = 2 h, where $\kappa = 32 \, \text{m}^2/\text{year}$.

3.3 Estimate the maximum temperature reached at the distance of z = D, 2 D, and 3 D from a dike intrusion using Fig. 3.6. The intrusion temperature is $T_1 = 1150 \, °C$, and the wall rock has an initial temperature of $T_2 = 150 \, °C$.

3.4 A spherical intrusion with radius R = 2.5 km and diffusivity $\kappa = 32 \, \text{m}^2/\text{year}$ cools to equilibrium. Estimate the time which is necessary to reach equilibrium within 10 % accuracy using Fig. 3.9.

3.5 A spherical intrusion of radius R cools from $T_1 = 750 \, °C$ to the temperature of the wall rock $T_2 = 100 \, °C$. Estimate the time needed to reach a temperature difference less than 50 °C, where R = 1 km and $\kappa = 32 \, \text{m}^2/\text{year}$.

4 Thermal State of the Earth's Interior

The age of the earth is estimated through radiometric dating methods to be 4.6 billion years. In the time since its origin, the earth has undergone a continuous thermal evolution. The thermal state which it has now reached can be estimated by direct and indirect means. However the difficulties in estimating temperature increase rapidly with depth. At present it is difficult to plot an accurate temperature map for crustal depths of only 10 km even in well-studied areas. Given the uncertainty in calculating the exact present day thermal state, one can imagine the difficulties and uncertainties in describing the thermal evolution of our planet through its 4.6 billion-year history.

The thermal state of the earth's interior must be described as a function of time and position. This function, in spite of all efforts, is relatively well known only near the earth's surface and at the present time. It is therefore appropriate to treat thermodynamics of the upper crust separately from that of the deeper interior.

4.1 Thermal State of the Upper Crust

A temperature measurement taken in a borehole is valid only for that particular place and time of measurement. Many environmental factors affect the value of the measurement. First, the subsurface temperature field can be altered by temperature variations at the surface such as the periodic daily and annual fluctuations or long-term climatic changes. Second, the temperature field is influenced by the morphology of the surface and geological structure of the adjacent crustal region. Third, water movement and tectonic events which are able to transport a large amount of heat through convection rather than conductive heat transfer in rock also have a marked influence. All these influencing factors, together with heat flow from the upper mantle and the lower crust, produce a thermal state in the crust which has measurable variation in time and space.

4.1.1 Influence of Climatic Variations on the Surface Temperature

The sun's energy, reaching the earth's surface, is very much greater than the thermal energy flowing upward through the earth's crust. Therefore it is solar energy which is the primary agent in determining the surface temperature of the earth. The radiation intensity, called the solar constant, has a value of $Q_s = 1.36 \, \text{kW/m}^2$. This value is valid for an average earth-sun distance, measured perpendicular to the sun rays and ignoring effects of the earth's atmosphere. The solar constant can be given at an accuracy of only a few percent, because the total energy spectrum is barely determinable and is probably also dependent on the sun's activity.

The solar spectrum can be divided in terms of its energy as follows [4.53]:

$$\begin{aligned}
&\text{visible light (wave lengths } 0.4-0.8\ \mu\text{m)} && 49\% \\
&\text{infrared light (wave lengths} \quad > 0.8\ \mu\text{m)} && 42\% \\
&\text{ultraviolet light (wave lengths} \quad < 0.4\ \mu\text{m)} && 9\%.
\end{aligned}$$

The total radiant energy (q_i) incident on a circular cross section having an earth radius of r is

$$q_i = \pi r^2 Q_s. \tag{4.1}$$

Of this incident energy about 30% is reflected back into space. This property is known as the albedo of the earth. The spectral composition of the reflected energy is different from that of the incident energy because the albedo is larger in the visible range than in the infrared part of the spectrum.

In the case of radiation energy, an equilibrium between incident and reflected energy, then thermal equilibrium (i.e. a steady state temperature condition) exists. Such is the case for the earth's surface. Since the radiating surface is the entire surface of the globe, the radiant energy is given by Stephan-Boltzmann's law as

$$q_r = 4\pi r^2 \sigma T^4 \tag{4.2}$$

with the Stephan-Boltzmann's constant

$$\sigma = 5.78 \cdot 10^{-8}\ \text{W}/(\text{m}^2\ {}^\circ\text{K}^4).$$

The equilibrium temperature (T) at the earth's surface can be calculated by combining (4.1) and (4.2), including effects of the earth's albedo. The result is

$$T = [(1 - A/100)\ Q_s/(4\sigma)]^{0.25} = 253\ {}^\circ\text{K} = -20\ {}^\circ\text{C}.$$

However the observed global average value of $T = 14\ {}^\circ\text{C}$ is substantially higher. The spectral shift of the short wave solar energy to the long wave heat radiation of the earth causes the difference. Multi-atom molecules of the atmosphere, especially water molecules, absorb long wavelength radiation and radiate it back to earth. Consequently comfortable surface temperature on earth that we experience must be attributed not only to primary solar radiation but also to secondary effects.

4.1.1.1 Diurnal and Annual Variations of the Surface Temperature

The surface temperature of the earth varies in conjunction with the movement of the planet. There are short period temperature changes having daily cycles, longer periodic variations of annual cycles and others with even longer periods. These periodic changes in temperatures penetrate to different depths in the earth. To calculate the penetration depth, one begins with the heat conduction equation (see Sect. 1.3) subjected to the boundary condition $T(z, 0) = 0$ and $T(0, t) = T_0 e^{i\omega t}$.

Such a boundary condition is appropriate for a periodic variation of surface temperature having frequency ω and amplitude T at the surface $z = 0$. The distur-

34

bance $T(z, t)$ at depth z and time t in material of thermal diffusivity κ can be expressed by an analytical solution with a Laplace transform [e.g. 4.81]:

$$T(z, t) = 0.5\, T_0 \exp(i\omega t)\, \{\exp(z\sqrt{i\omega/\kappa})\, [1 - \mathrm{erf}(z/2\sqrt{\kappa t} + \sqrt{i\omega t})]$$
$$+ \exp(-z\sqrt{i\omega/\kappa})\, [1 - \mathrm{erf}(z/2\sqrt{\kappa t} - \sqrt{i\omega t})]\}. \tag{4.3}$$

This expression can be considerably simplified if one considers the temperature variation only after a long time. Then the initial "transient phenomenon" is negligible and the solution is periodic with time and exponentially decreasing with depth:

$$T(z, t) = T_0 \exp(-z\sqrt{\omega/(2\kappa)})\, \cos(\omega t - z\sqrt{\omega/(2\kappa)}). \tag{4.4}$$

The maximum temperature variation caused by the periodic perturbation at the surface decreases exponentially with depth

$$T_{max}(z) = \pm\, T_0 \exp(-z\sqrt{\omega/(2\kappa)}). \tag{4.5}$$

One can thus define a penetration depth (z^*) at which the maximum temperature variation $T_{max}(z)$ decreases to T_0/e of its value at the surface. This gives

$$z^* = \sqrt{2\kappa/\omega}. \tag{4.6}$$

The temperature amplitude T_0 is strongly dependent on meteorological factors and geographic location. Furthermore the daily cycle is subjected to many more disturbing influences than annual cycles.

In order to measure the depth of penetration (z^*) correctly, there should be several days of consistent weather conditions. Not only must the "transient phenomenon" fade away but the moisture content of the soil must reach a stable state because it influences the thermal diffusivity (κ).

A computed daily temperature cycle is shown in Fig. 4.1. The ratio $T(z, t)/T_0$ is plotted against depth z with time t as a parameter. This shows the penetration

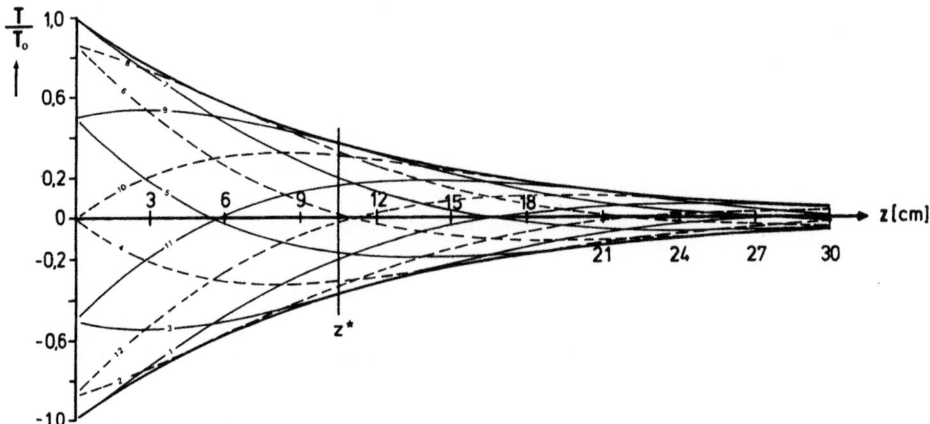

Fig. 4.1. Penetration of a sinusoidal daily temperature wave into the earth. Wave amplitude at the surface is T/T_0. Individual curves represent temperatures at different times. To obtain the time multiply the curve parameter by the factor 2

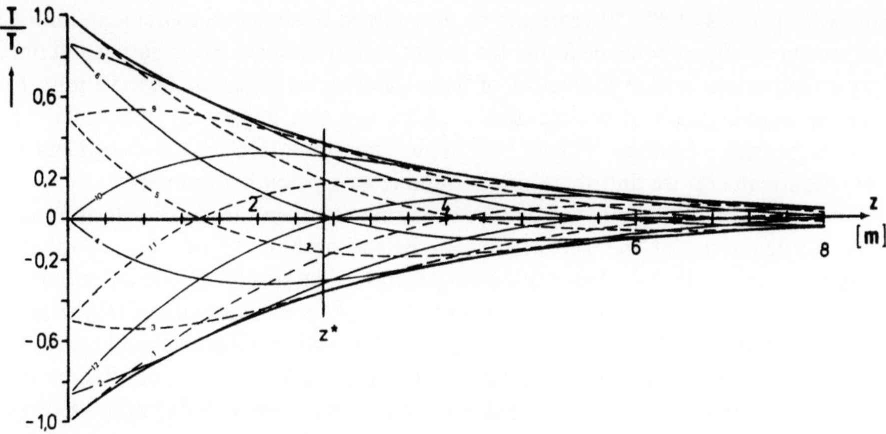

Fig. 4.2. Penetration of a sinusoidal annual temperature wave into the earth. Wave amplitude at the surface is T/T_0. Curve parameter is time (months)

depth of the daily temperature wave is $z^* = 10.5$ cm for a thermal diffusivity $\kappa = 4 \cdot 10^{-7}\,\mathrm{m^2/s}$. The envelope of the isochron curves agrees with the maximum values calculated from equation (4.5).

The surface amplitude of the annual temperature wave in middle Europe is $T_0 \approx 8$ to 9 °C. It is a little lower along the coastal areas and a little higher away from the sea. In the Federal Republic of Germany the mean annual temperature at the earth's surface ($\bar{T}$) is between about 8 and 11 °C. The seasonal variation of the temperature has greater penetration because of its longer duration.

Fig. 4.2 shows the temperature ratio $T(z, t)/T_0$ versus depth z for an annual wave. Time t in months is the curve parameter for this representation. A thermal diffusivity of $\kappa = 7 \cdot 10^{-7}\,\mathrm{m^2/s}$ is assumed. In the case of this annual wave the penetration depth is $z^* = 2.7$ m. The envelope which describes the maximum amplitude of the isochron curves is again computed from equation (4.5).

This annual temperature wave penetrates to a much greater depth than does the daily wave. Its influence at $z = 30$ m depth is still about $T = 10^{-3}$ °C. When taking a temperature measurement in a borehole, one should consider the influence of annual temperature wave to much greater depths than the penetration depth z^*.

4.1.1.2 Long-Term Temperature Variations

The surface temperature of the earth has changed many times in the course of its geological history. It is known that there have been widespread glacial events in the Precambriam, in Permo-Carboniferous time and in the Quaternary. These global glacial periods are separated in time by similar intervals of a few hundred million years.

Long-period fluctuations can be recognized unmistakably only in much shorter time cycles, for example those fluctuations which have about three to four times the 11-year sunspot cycle. The temperature amplitudes of such variations are indeed very small and barely influence the subsurface temperature field. While temperature varia-

36

tions with periods of 30–50 years can be determined by direct measurement, indirect
temperature evidence must be found for longer period changes. Fossil fauna and flora
play an important role as indicators of temperatures. Lithogenetic observations, for
example examinations of sedimentation cycles are also important indicators. The
ratio of oxygen – isotope ^{18}O to ^{16}O in the limestone ($CaCO_3$) is dependent on
formation temperature and therefore is suitable as a paleothermometer.

With some reliability one can deduce the annual temperature for Western Europe
since the beginning of the Tertiary. From Paleocene time on one can establish a
continual decrease in the mean annual temperature. The steady climate deterioration
in Tertiary time is a worldwide phenomena [4.66]. As a measure of the temperature
decline a botanical criteria is given. The assemblages of tree flora depend to a great
extent on the mean annual temperature and the amplitude of the annual fluctuation.
Since Tertiary time the terrestrial flora is somewhat the same as today and the recent
flora serves as a calibration of the paleothermometer. The assemblage of deciduous
trees is especially sensitive to temperature. Often the ratio of smooth edge leaves (e.g.
willow) to the total number of leaves of trees and bushes (such as willow, linden,
hazelnut etc.) is used as an indicator of temperature. The presence of smooth edge
leaves lets one conclude that for the Paleocene (about 55 million years ago) there was
a mean annual temperature of 25 °C in West Europe, which decreased steadily till the
Pliocene to 10–15 °C [4.22].

With the start of the Quaternary the temperature dropped sharply to the Quaternary ice age minimum, interrupted by short warm periods. Even though for the
Quaternary at present complete research information is available regarding the plant
community of individual regions, [e.g. 4.26] a general answer to the annual average
temperatures is not possible. In Fig. 4.3 the probable temperature variations on the
earth's surface since early Tertiary time are shown on a logarithmic time scale.
Thereby the temperature changes in the ice age could be assumed to be quasi periodic.
However, since the Early Tertiary the average temperatures are subjected to yet

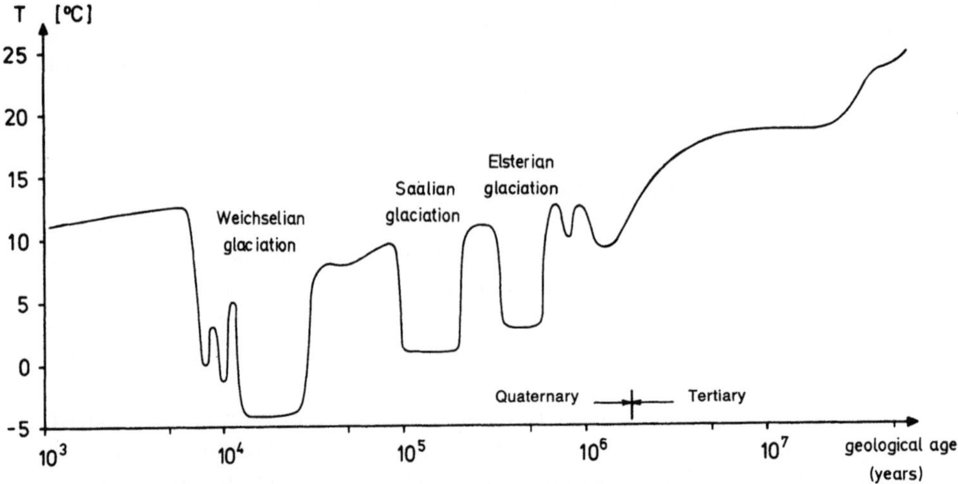

Fig. 4.3. Surface temperature in NW Europe since Tertiary time after [4.22, 4.45, 4.70]

another long-term temperature decrease. It is not possible at this time to prove if this trend was part of a temperature fluctuation which lasted a few hundred million years.

4.1.2 Topographic Influence on the Subsurface Temperature Field

In the case of a flat horizontal surface at constant temperature and overlying a homogeneous medium, isotherms underground lie parallel to the surface. No horizontal gradient of the temperature can exist. However a change in the temperature field is caused by mountains and valleys through their morphology. In this case the isotherms do not lie parallel to the earth's surface or to each other bacause of the topography. A horizontal temperature gradient can now exist. The deviations are such that the temperature gradient under a hill is smaller than under a flat surface. Under a valley the gradient is greater than under the flat surrounding region.

The topographic influence can be calculated using Laplace's equation. If radioactive heat production cannot be ignored, then Poisson's equation serves for the description of the isotherms. For such a problem, which has to be solved in at least two dimensions, there are various approximation methods to the analytical solution [e.g. 4.5, 4.9, 4.44, 4.47, 4.48]. With the use of computers solutions to the appropriate differential equation can be found using numerical methods [e.g. 4.30].

A topographic thermal disturbance in the earth also occurs because air temperature decreases with increasing elevation. Thus surface temperatures are reduced in correspondence to the elevation ($z < 0$) above sea level. In the Alps this temperature gradient $[(dT/dz)_B]$ is approximately $(dT/dz)_B = 4.5\,^\circ C/km$. For the calculation of the undisturbed temperature gradient prevailing for the same heat flow under a flat surface, stationary conditions are used. In other words, one assumes that mountains with their present day topography have existed unchanged for a long time. The deeper under the surface the point of correction lies, the more accurate the stationary analysis becomes.

The topographic correction [4.9, 4.44] is completed in two steps. First, the thermal effect of the terrain which has to be corrected for is projected on to a flat plane. The resulting temperature distribution on the plane is equivalent to the effect of the uneven topography. An approximation at this stage consists of using the uncorrected measured temperature gradient (dT/dz) to calculate the deviation from a constant temperature on the reference plane

$$T_B = h[dT/dz - (dT/dz)_B].\tag{4.7}$$

These temperatures T_B on the plane are determined as the average values within concentric rings around the point P, and depend on the average elevation h determined from a topographic map (see Fig. 4.4).

This previously determined temperature distribution is now used in the second step. Poisson's integral of potential theory is calculated for the special case such that the required temperature gradient $(dT/dz)_c$ at point P of the half space satisfies Laplace's equation with boundary function $T_B(r)$:

$$(dT/dz)_c = \int_0^\infty (1 - 2z_0^2/r^2)/(1 + z_0^2/r^2)^{5/2}\,\frac{T_B(r)}{r^2}\,dr.\tag{4.8}$$

38

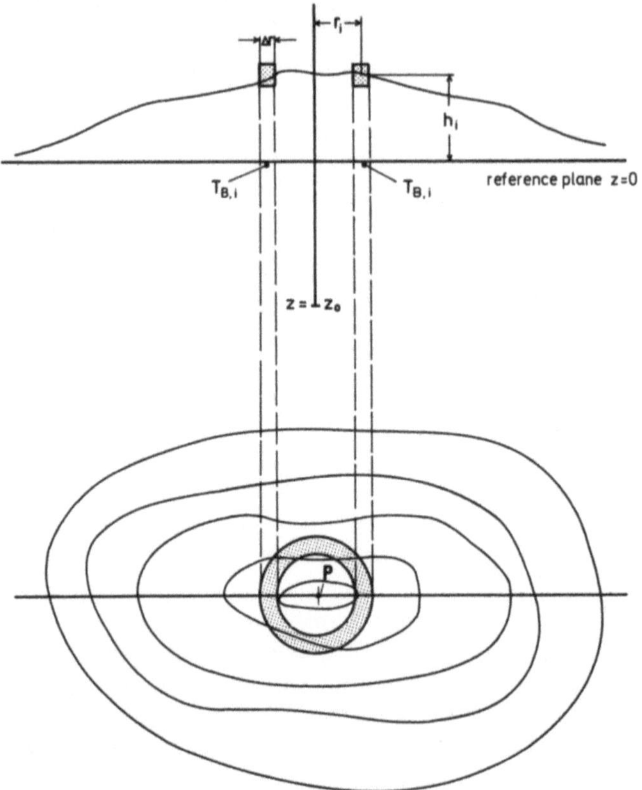

Fig. 4.4. Model of topographic correction after [4.9, 4.44]

The integral corresponding to Fig. 4.4 is solved numerically

$$(dT/dz)_c \approx \sum_{i=1}^{n} (1 - 2z_0^2/r_i^2)/(1 + z_0^2/r_i^2)^{5/2} \frac{T_{B,i}}{r_i^2} \Delta r$$

$$= \sum_{i=1}^{n} (1 - 2z_0^2/r_i^2)/(1 + z_0^2/r_i^2)^{5/2} \frac{h_i \Delta r_i}{r_i^2} \left\{ \frac{dT}{dz} - \left(\frac{dT}{dz}\right)_B \right\}. \tag{4.9}$$

The corrected temperature gradient $(dT/dz)_0$ at the point P is given by:

$$(dT/dz)_0 = dT/dz - (dT/dz)_c. \tag{4.10}$$

The same methods can be applied to determine the correction beneath a valley in which case $h < 0$. If the topographic relief is very large then this approximation can lead to considerable errors [4.24].

The influence of uplift and erosion in geologic time also can be calculated to correct the temperature gradients [4.5]. However, neither the rates of uplift nor erosion are well known. Furthermore both are not constant through geologic time. This correction is often ignored. Other effects which are likewise often overlooked are local variations of thermal conductivity, their anisotropy, secular climatic variation, radioactive heat production and the movement of ground water. With a few examples from the Central Alps, one can illustrate corrections for various influences on the temperatures in boreholes and tunnels [4.6].

4.1.3 Changes in the Temperature Field Caused by Water Movement

A temperature disturbance spreads much more slowly in the underground through pure conduction than if there is additional convective heat transport associated with a migrating fluid phase. Water movement usually leads to local disturbances of the geothermal field. Regional disturbances from water movement may also occur but are more difficult to demonstrate. Regional effects can be confused with effects stemming from the deeper earth which are often unknown. Large area disturbances of thermal fields can, for example, be observed in sedimentary basins.

Two simple types of water movement are distinguishable by their thermal effects. First, when meteoric water percolates from the surface to the subsurface, it absorbs heat and warms up. Because it must extract heat from its surroundings it decreases subsurface temperatures. Second, and in contrast, hydrothermal water discharging from deeper layers of the crust warms up the rocks in passing through and cools itself. Usually the second case is local, while in the first the influence is of a regional nature.

An example of a sediment basin with a large area flow system is the Pannonian Basin (Fig. 4.5). Surface water penetrates into the permeable sedimentary rock at the edge of the foothills of the Carpathians and migrates within these sedimentary layers towards a central basin. Here they again come to the surface as thermal springs. This circulation pattern causes positive deviation of the temperature field of the upper crust in the discharge zone and negative deviation in the recharge zone [4.80]. Flow systems, connected with positive and negative thermal anomalies of a smaller scale are observed in the Lower Rhine Embayment [4.3].

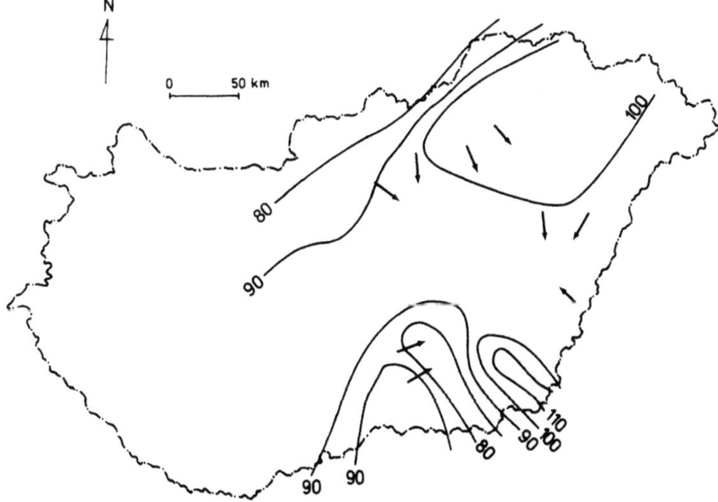

Fig. 4.5. Migration of pore water (→) of the Great Hungarian Plain after [4.80] and lines of equal heat flow density [mW/m^2] after [4.39]

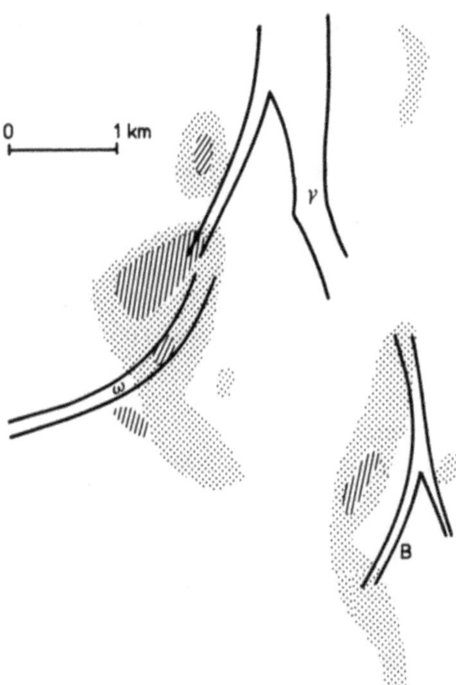

Fig. 4.6. Temperature anomaly of Landau/ Pfalz showing elevated ▓▓▓ and maximum ///////, values of the anomaly in relationship to the position of fracture zones γ, ω, B after Doebl in [4.13]

Thermal springs are much more recognizable as thermal anomalies and have been utilized since ancient times. The temperature fields produced by these are in most cases only locally exploitable. However individual springs can be a visible indicator of a larger deep-lying thermal source. In the tectonically active Rhinegraben, deep water can form an effective large discharge system [4.13, 4.85, 4.87] which explains the anomaly at Landau/Pfalz. Fig. 4.6 shows clearly that in the Landau area high local temperature anomalies could be detected in the fault zones. A simple model can show how such a thermal anomaly can arise and the order of magnitude the corresponding physical parameters must have.

In a convection model, [4.13] one assumes that local heat sources in the deep subsurface lead to horizontal density gradients caused by temperature differences. These in turn produce hydrostatic pressure gradients in communicating porewater. With sufficient permeability in the overlying rocks a condition can form in which water from a large reservoir is locally heated and rises in a relatively narrow vent. The critical parameter in such a situation is the vertical permeability in the discharge region.

With the help of Darcy's equation and the size of the anomalous heat flow density the necessary permeability can be calculated. It is assumed that the flow resistance for the lateral recharge flux is small compared to that for local discharge. One can then write for the filtration velocity v_f:

$$v_f = \frac{\xi}{\eta}\frac{\Delta p}{h} \tag{4.11}$$

with permeability ξ, viscosity η, pressure difference Δp and the vertical discharge height h. The heat flow density caused by the migrating fluid amounts to

$$Q_W = \varrho \, \sigma \, v_f \, \Delta \bar{T} = \frac{\varrho \, \sigma}{\eta} \, \xi \, \frac{\Delta p}{h} \, \Delta \bar{T}. \tag{4.12}$$

In this equation ϱ is the density of water, σ the specific heat and $\Delta \bar{T}$ the average temperature difference with the surroundings. The water parameters are given at normal conditions and thus yield an estimation on the unfavorable side.

For Landau the region of maximum thermal anomaly has a 50 °C/km higher temperature gradient compared to the background region. The Landau anomaly runs to a depth of h = 2000 m to an average of $\Delta \bar{T} = 50$ °C hotter water column compared with it's surroundings. This temperature difference causes a density difference of $\Delta \varrho = 0.01 \text{ g/cm}^3$, and a corresponding pressure difference of $\Delta p = \Delta \varrho \, gh = 2 \cdot 10^5 \text{ N/m}^2$ forms.

In order to explain an anomaly of 60 mW/m^2 one requires an average permeability of $\xi = 3 \cdot 10^{-15} \text{ m}^2 \approx 3$ millidarcy throughout the depth interval. Such permeability is certainly reached.

A simple model for the estimation of the temperature distribution in the vicinity of a fault zone, in which water discharges from the depth z_0 with a temperature $T(x, 0) = T_2$ is shown in Fig. 4.7.

The model is applicable for a small layer thickness for which the discharging water is only slightly cooled as it reaches the surface z = 0 with temperature $T(x, 0) = T_1$.

Laplace's equation

$$\frac{\partial^2 \hat{T}}{\partial \hat{x}^2} + \frac{\partial^2 \hat{T}}{\partial \hat{z}^2} = 0 \tag{4.13}$$

taken with the boundary condition

$$\hat{T}(0, \hat{z}) = 1$$
$$\hat{T}(\hat{x}, 0) = 0$$
$$\hat{T}(\hat{x}, 1) = 1$$

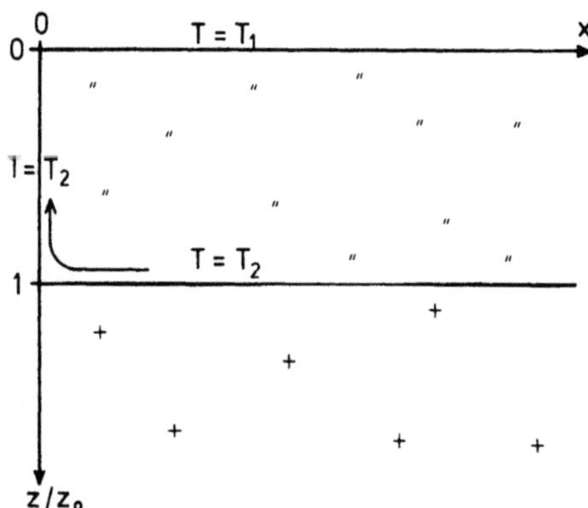

Fig. 4.7. Model for estimating the temperature distribution near a fracture zone which acts as a conduit for ascending thermal water after [4.89]

yields the solution [4.89]:

$$\hat{T}(\hat{x}, \hat{z}) = \hat{z} + \frac{2}{\pi}\left[\arctan\frac{\sin(\pi\hat{z})}{\sinh(\pi\hat{x})} - \arctan\frac{\sin(\pi\hat{z})}{\exp(\pi\hat{x}) + \cos(\pi\hat{z})}\right]$$

with

$$\hat{T}(\hat{x}, \hat{z}) = \frac{T(x, z) - T_1}{T_2} \quad \text{and} \quad \hat{z} = z/z_0, \ \hat{x} = x/z_0. \tag{4.14}$$

From equation (4.14) one can derive the vertical temperature gradient

$$\frac{\partial\hat{T}}{\partial\hat{z}} = 1 + \frac{2\sinh(\pi\hat{x}) - \cos(\pi\hat{z})}{\sinh^2(\pi\hat{x}) + \sin^2(\pi\hat{z})} - \frac{2\exp(\pi\hat{x})\cos(\pi\hat{z}) + 2}{\exp(\pi\hat{x}) + \cos(\pi\hat{z})^2 + \sin^2(\pi\hat{z})}. \tag{4.15}$$

These estimations of the maximum possible disturbances are based on stationary models, which are only applicable to limited crustal regions. Time dependent models demand a much greater mathematical effort. This must not be justified for larger crustal regions, because the water discharge locations in general are not fully known. Furthermore many of the systems cannot be explained fully with such a simple model concept. As a result time-dependent observations are as a rule only of local and site-specific significance.

In the spreading of warm water in the subsurface, for example industrial waste water injected into a borehole, the groundwater and groundwater medium warms up. Then one must determine the solution to the following differential equation:

$$\frac{\partial T}{\partial t} = -c_w \varrho_w v_f/(c\varrho)\, \nabla T + (D_0 + \kappa)\, \Delta T. \tag{4.16}$$

The time-dependent temperature T is given in terms of c_w, c the specific heat of water and water rock system respectively ϱ_w, ϱ the density, v_f the filteration velocity, D_0 the diffusion constant and κ the thermal diffusivity. The considered 3-layer case [4.86] is difficult to solve analytically. For the numerical solution, the differential equation is substituted with a difference equation which can be solved by various methods [e.g. 4.76]. In this example [4.86] the given quantities are: the temperature of the injected water $T = 45\,°C$ and its flow $v = 6.7\,m^3/day$. The filteration velocity of the groundwater is $v_f = 0.23$ cm/day and the ratio $\frac{c_w \varrho_w}{c\varrho} = 0.9$. The cover is $z = 0.9$ m thick and the permeable sand layer over the clay halfspace $z = 2.7$ m thick. The parameters $(D_0 + \kappa)$ for each layer are estimated at 15, 20, and 10 m^2/day respectively. The theoretical model and the experiment show after 64 days that in the distance of about 7 m from the injection borehole the temperature has already decreased by a factor of 1/e (about 36.8%).

4.1.4 Temperature Fields in Different Types of Geological Structures

Bounding surfaces separating two rock types can take several forms. If in various units the thermal properties are identical, the contact zones produce no disturbance in the temperature field. With a constant heat flow density from the earth's interior, the temperature gradient remains the same everywhere. If, however, rock bodies of differing thermal properties are in contact, then in the area of contact a disturbance occurs in the temperature field and in the thermal gradient field. The greater the thermal conductivity contrast, the greater the interference.

An estimation of the deviations in the temperature field of small geological units are dealt with in the following examples. The assumed stationary conditions are given in Section 3.1.

In a fault zone, rocks of varying properties often border on each other. Magmatic events can also lead to boundary surfaces across which the material properties of the intrusive rock and the country rock differ.

In the first example (Fig. 4.8) two bodies form a contact, which is perpendicular to the ground surface. For a constant heat flow density Q from the interior of the earth, differing temperature gradients are produced in both bodies 1 and 2 depending on their different thermal conductivities:

Body (1): $(dT/dz)_1 = Q/K_1$

Body (2): $(dT/dz)_2 = Q/K_2$.

At large distances from the contact the horizontal difference in temperature at depth z is given by

$$\Delta T = T_2(z) - T_1(z) = [(dT/dz)_2 - (dT/dz)_1]z = (dT/dz)_1 (K_1/K_2 - 1)z. \quad (4.17)$$

Body (2) represents the lower thermal conductivity $(K_2 < K_1)$. At the boundary surface the temperature $T_b(z)$ is given by

$$T_b(z) = T_0 + \left(\frac{dT}{dz}\right)_1 \left[1 + \left(\frac{K_1}{K_2} - 1\right)\frac{V}{1 + V}\right]z \quad (4.18)$$

in which

$$V = \sqrt{\frac{K_1 c_1 \varrho_1}{K_2 c_2 \varrho_2}} = \frac{K_1}{K_2}\sqrt{\frac{\kappa_2}{\kappa_1}}.$$

The temperature gradient at the boundary surface equals

$$dT_b/dz = (dT/dz)_1 (1 + (K_1/K_2 - 1) V/(1 + V)). \quad (4.19)$$

Differences in the thermal fields increase with increasing difference in the thermal properties of the two bodies. Large differences exist, for example, between sandstone and coal or rocksalt and sandstone.

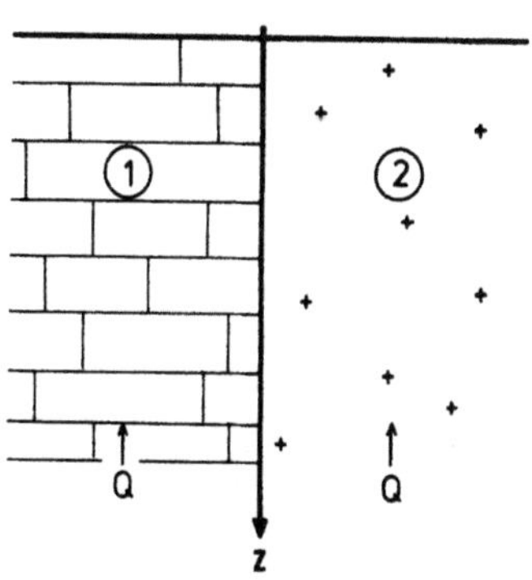

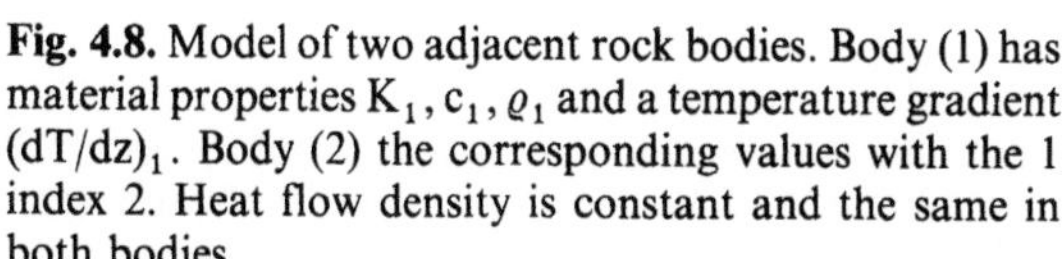

Fig. 4.8. Model of two adjacent rock bodies. Body (1) has material properties K_1, c_1, ϱ_1 and a temperature gradient $(dT/dz)_1$. Body (2) the corresponding values with the index 2. Heat flow density is constant and the same in both bodies

44

With the material properties given in Table 2.1 $\left(\text{rocksalt } K_1 = 5.5 \dfrac{W}{m\,^\circ K},\right.$ sand-

stone $\left. K_2 = 3.2 \dfrac{W}{m\,^\circ K}\right)$ the relationship between temperature gradient is

$$\left(\frac{dT}{dz}\right)_2 = \frac{K_1}{K_2}\left(\frac{dT}{dz}\right)_1 = 1.7\left(\frac{dT}{dz}\right)_1.$$

In the case of constant heat flow density, the temperature gradient in sandstone is 1.7 times higher than in rocksalt.

Consider now the case of horizontal layering of n materials with different thermal properties K_i (Fig. 4.9) and thicknesses h_i and with a constant heat flow density Q. One can calculate an average thermal conductivity K_m and an average temperature gradient $(dT/dz)_m$ within the layer thickness H as follows:

$$Q = K_m(dT/dz)_m = K_m Q/H \sum_{i=1}^{n} \frac{h_i}{K_i} \tag{4.20a}$$

$$\frac{1}{K_m} = \frac{1}{H}\sum_{i=1}^{n} \frac{h_i}{K_i} \tag{4.20b}$$

$$(dT/dz)_m = Q/K_m. \tag{4.20c}$$

For example, in the coal seams of the Ruhr district where up to one meter thick seams of coal ($K_2 = 0.26$ W/m $^\circ$K) are interlayered between 10 meter layers of sandstone ($K_1 = 3.2$ W/m $^\circ$K) an average thermal conductivity would be

$$K_m = (10 + 1)(1/0.26 + 10/3.2)^{-1} = 1.6 \text{ W/m } ^\circ K.$$

For calculations of temperatures below this interlayering, one can count on an average temperature gradient only within the overlying layers.

The temperature in region (4) of Fig. 4.9 is given by:

$$T(z) = T_0 + Q\sum_{i=1}^{3} h_i/K_i + \left(z - \sum_{i=1}^{3} h_i\right)\left(\frac{dT}{dz}\right)_4 \tag{4.21}$$

Fig. 4.10 shows a combination of layering and faulting.

A comparison of the temperature gradients in the Ruhr coal district and the unlayered sandstone shows how large the thermal anomaly can become between two

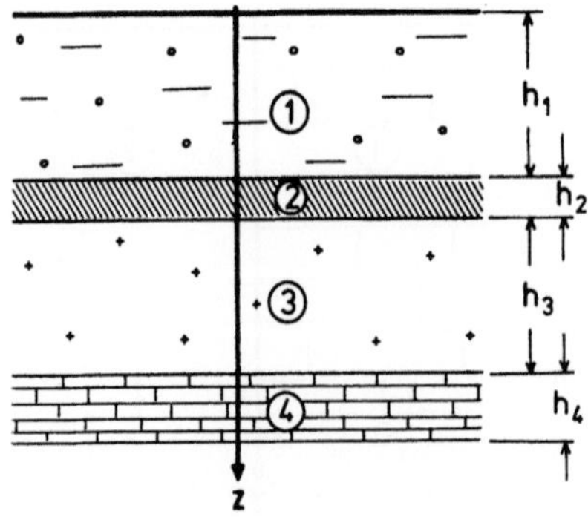

Fig. 4.9. Model for horizontal layering of n rock layers, each having its own thickness h_n and material properties (K_n, c_n, ϱ_n)

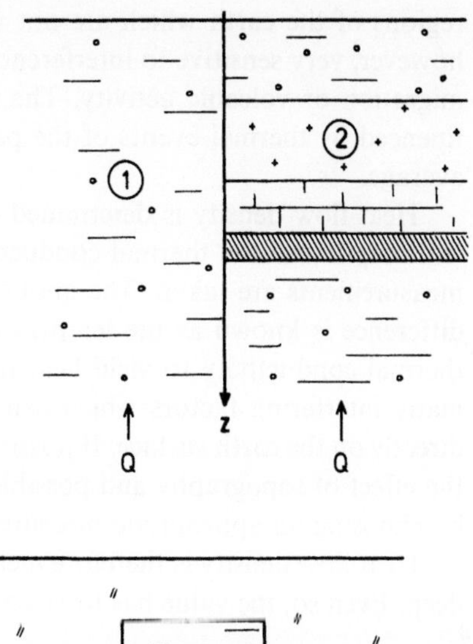

Fig. 4.10. Model for a combination of faulting and horizontal layering of different rock types

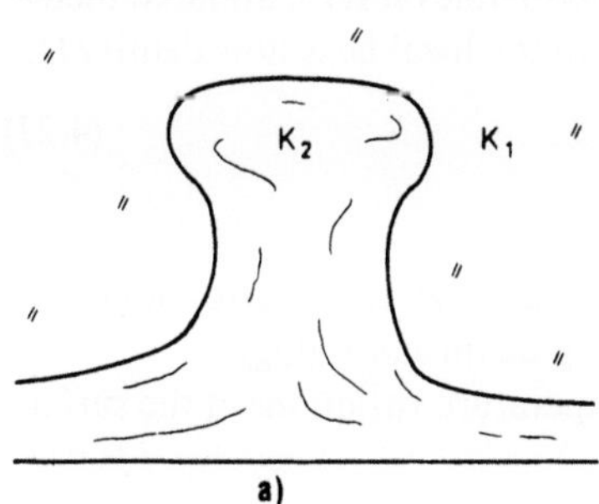

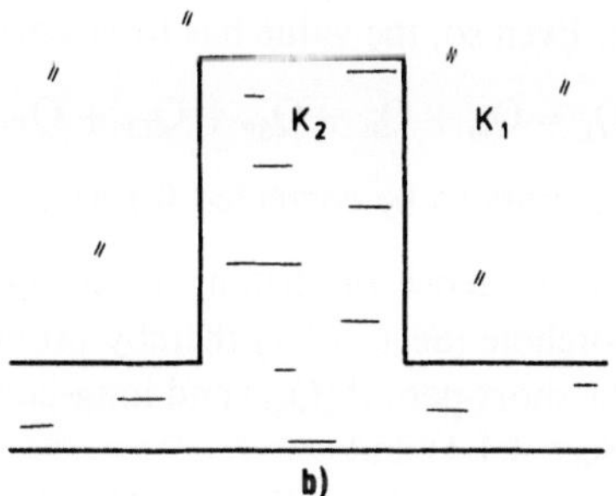

a)

b)

Fig. 4.11 a, b. Structure of a salt dome (**a**) and simplified model (**b**). Thermal conductivity of the country rock is (K_1) and salt is (K_2)

different types of rock complexes. The sandstone has a thermal conductivity $K_1 = 3.2$ W/m °K and the layered complex which includes the coal an average conductivity of $K_m = 1.6$ W/m °K. The temperature gradient (dT/dz) is thus

$$\left(\frac{dT}{dz}\right)_m = \frac{K_1}{K_m}\left(\frac{dT}{dz}\right)_1 = 2\left(\frac{dT}{dz}\right)_1$$

twice as large as in the sandstone under conditions of equal heat flow density.

In geometrically complicated structures, as for example, in salt domes (Fig. 4.11 a) one can calculate the thermal field only by numerical methods [4.27, 4.54]. A first approximation can be obtained, if the geometry of the salt dome is approximated by a simple body (Fig. 4.11 b). For the central region, and also the area further from the salt dome, the temperature gradient can be estimated as in the case of horizontal stratification.

4.1.5 Terrestrial Heat Flow Density

The heat flow density from the interior of the earth is that thermal quantity which allows one to make the first approximation of the thermal condition of the deeper regions of the earth which are not accessible for temperature measurements. It is, however, very sensitive to interference caused by crustal uplift and subsidence, water migration or volcanic activity. The magnitude of the heat flow density can be influenced by thermal events of the past. Thus it has to be considered as a transient average.

Heat flow density is determined by measuring the vertical temperature increase with depth and the thermal conductivity of the rock within which the temperature measurements are taken. The quotient of the temperature difference and the depth difference is known as the temperature gradient. The gradient is multiplied by the thermal conductivity to yield heat flow density (compare Sect. 1.2). Because of the many interfering factors which can occur, heat flow density cannot be measured directly on the earth surface. If possible, periodic temperature changes on the surface, the effect of topography and possible ground water disturbances should be avoided by choosing an appropriate measurement depth.

Heat flow density is therefore measured (Q_m) in boreholes several hundred meters deep. Even so, the value has to be corrected to obtain the local heat flow density Q_L:

$$Q_L = Q_m + Q_B + Q_{SP} + Q_{LP} + Q_T. \tag{4.22}$$

The factors to be corrected for are:

1. the influence of drilling fluid which cools or heats the rock surrounding the borehole (Sect. 5.2.1) thereby producing a drilling disturbance (Q_B),
2. the short-period (Q_{SP}) and long-period (Q_{LP}) temperature variations at the surface (Sect. 4.1.1) and
3. the topographic influence (Q_T) (Sect. 4.1.2).

The corrected local heat flow density can be considered as a thermal anomaly, if it deviates from a regional average value or reference value.

It is much simpler to take heat flow density measurements at the bottom of oceans and seas because drill holes are unnecessary. Measurement probes penetrate the soft sediments under their on weight far enough for the temperature gradient and in-situ-thermal conductivity of the sediments to be determined [4.31, 4.32]. Because measurements are made under several hundreds to thousands of meter of water, the correction of the short period temperature changes can be neglected [4.50, 4.68, 4.84].

4.1.5.1 Regional Variation of Heat Flow Density

The average continental heat flow density is smaller than the average oceanic heat flow density. Mean values are reported for continents as $55 \pm 5 \, mW/m^2$ and for oceanic crust as $95 \pm 10 \, mW/m^2$ [4.20]. Yet the heat flow density on the continents is made up of parts quite different from those of the ocean floor. The phenomenon of forming new crust along the mid ocean ridges contributes, due to the convective heat transport of the warmer rising crustal material, to the average oceanic value. The greater oceanic heat flow is due to the significant hydrothermal contribution in young oceanic crust. In order to estimate the thermal condition of the earth's interior, the correction of the measured data plays an essential role for interpretation. Standards for this procedure should be used [4.33]. The contribution from radioactive heat production is minimal, because the 5 km thick basalt layer of the oceanic crust contains no large concentrations of uranium, thorium or potassium. In addition to these two effects, hydrothermal circulation of water in the oceanic crust results in large local heat flow variations.

Oceanic heat flow density is substantially influenced by new crust formation. The heat flow density is highest near mid oceanic ridges and can have a magnitude as large as $Q = 300 \, \text{mW/m}^2$ [4.74]. Perpendicular to the ridge heat flow density decreases exponentially and falls to a value of $Q = 40$ to $45 \, \text{mW/m}^2$ at a distance of about 8000 km. These lower values no longer include any heat contribution from the convective upwelling of basalt from the earth's interior to the surface of the ocean floor. The heat flow density of $Q = 40$ to $45 \, \text{mW/m}^2$ consists of one part which reaches the surface through heat flow from the upper mantle, and a second part which results in the crust through the decay of the unstable isotopes of the uranium, thorium and potassium. The latter is very slight, so that the heat flow density out of the upper mantle of the edge of oceans amounts to about $Q_m = 40 \, \text{mW/m}^2$ [4.59]. A typical distribution of heat flow density in the oceanic crust is shown in Fig. 4.12.

Oceanic heat flow density is also correlated with the topography of the ocean floor. A thermal model of sea floor spreading in which the lithosphere is about 100 km thick and spreads evenly from the ridge corresponds with the observed heat flow data [see 4.74].

The regional distribution of heat flow on the continents shows a many-faceted picture. Two crustal types can be qualitatively distinguished by the heat flow density: first the old shields, or the Precambrian regions and secondly the "normal type" crust

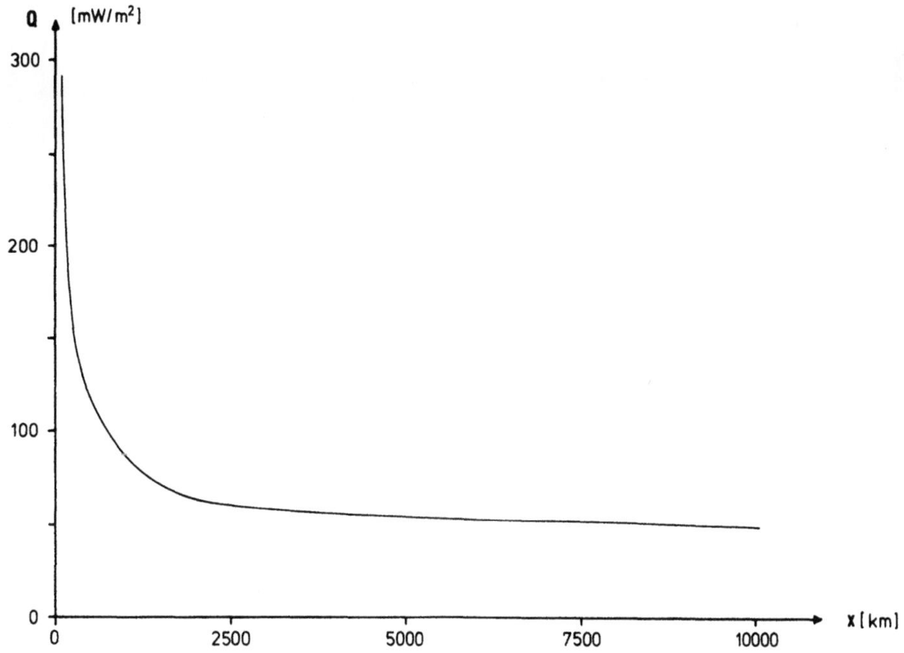

Fig. 4.12. Average heat flow density on the ocean floor showing its dependence on the distance from mid oceanic ridges after [4.73]

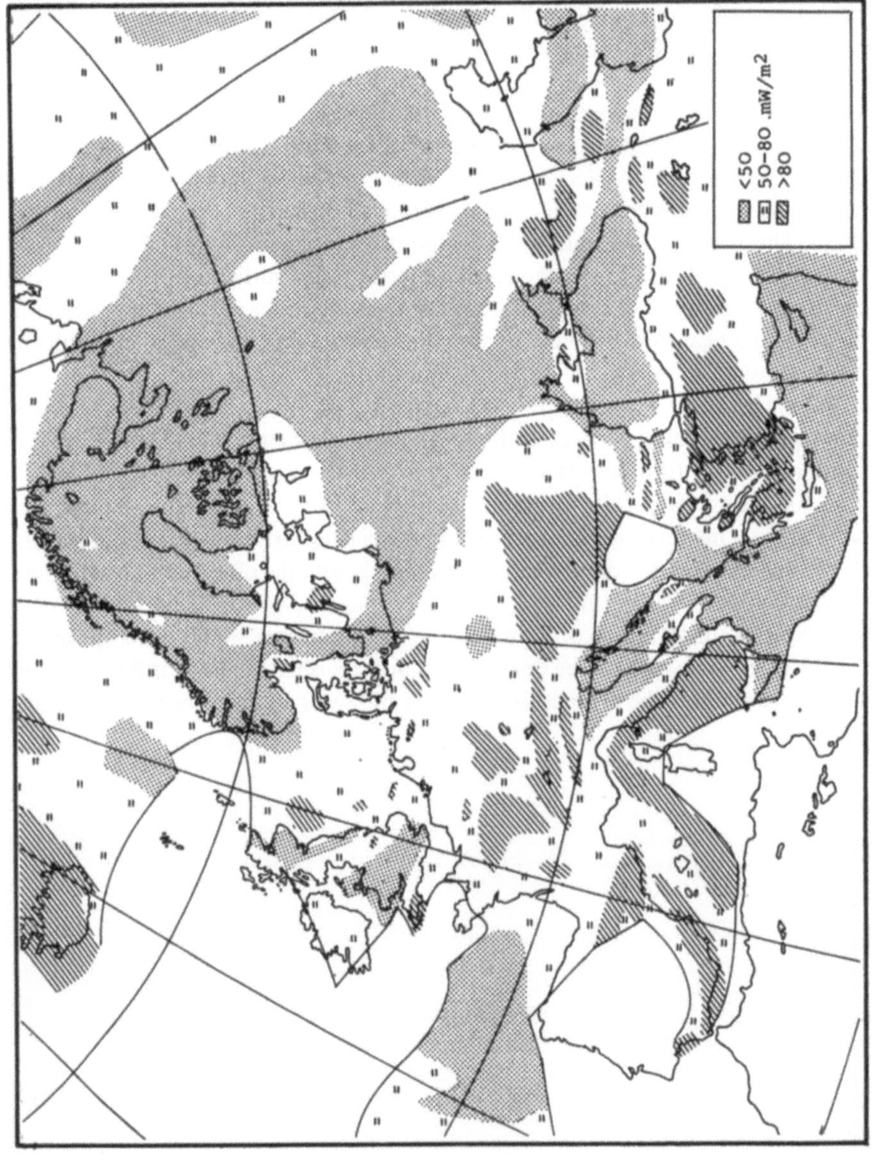

Fig. 4.13. Distribution of heat flow density in Europe after [4.15, 4.40, 4.41]

[4.7], which includes the Cambrium and younger regions. In the old shields the crust-mantle boundary (Mohorovičić discontinuity) is found at a depth of about 40 km and in regions of normal type crust at a depth of about 30 km when the deep-seated roots of mountains are excluded. In the old shield areas heat flow density is generally lower than in geologically younger regions.

An overview of the characteristic distribution of heat flow density on continents is shown by the heat flow density map of Europe (Fig. 4.13) [4.40, 4.41].

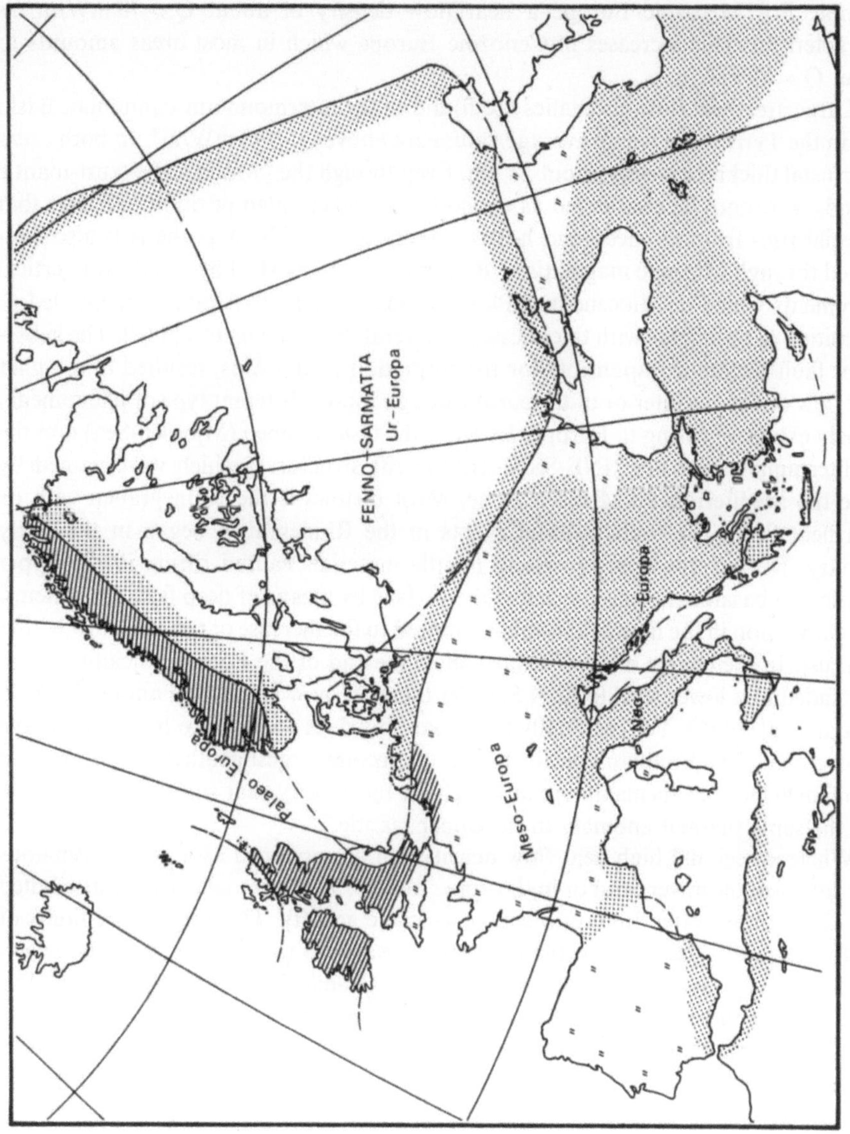

Fig. 4.14. Tectonic consolidation regions of Europe after [4.67]

Tectonic units are essentially reflected in the heat flow density pattern. Archaen-Europe with the Baltic and Ukrainian shield and the East European platform make up a large region with very low heat flow density ($Q < 50$ mW/m^2). This is bounded by the Urals in the East and Tornquist line in the West. Paleozoic Europe consists of Caledonian structures especially along the Norwegian coast and in Scotland showing a somewhat higher heat flow density ($50 \leq Q \leq 60$ mW/m^2). The entire region of Archean-Europe is bounded by these zones of slightly higher heat flow density. The

50

Bohemian Massif (Variscian) has a distinctly lower heat flow density than Mesozoic Europe. For Mesozoic Europe a heat flow density of about $Q = 70 \, mW/m^2$ is characteristic. This increases in Cenozoic Europe which in most areas amounts to about $Q = 80 \, mW/m^2$.

Large area heat flow anomalies are found in the intermountain Pannonian Basin and in the Tyrrhenian where average values are above $Q = 90 \, mW/m^2$. In both areas the crustal thickness is only about 25 km. Even though the "Moho", the crust-mantle boundary, cannot be considered as an isotherm, the elevated position indicates that material rises from the deep and hotter asthenosphere. The hypothesis is also confirmed through Miocene magmatism. Both regions are marked by significant vertical movements which in Miocene and Pliocene times, after the Alpine orogeny, led to deposition of sediments with thicknesses of several thousand meters [4.67]. The subsequent fault tectonics responsible for the formation of the Alps, resulted in a higher heat flow out of the interior of the earth through totally different type of phenomena, namely extensive rifting in Europe. Between the Oslo Graben (Mjoesen-Sea) and the Mediterranian there are NNE-SSW-striking rift structures, which were named by Stille the Mediterranean-Mjoesen Zone. Most distinct is the Rhinegraben with its high heat flow. The fault tectonic events in the Rhinegraben began in the Early Tertiary. During this time upwelling mantle materials caused rifting in the upper mantle and basaltic materials reached the surface by means of deep fracture systems. The convection in the upper mantle contributed to the increase of the heat flow within the crust. In the course of the Tertiary, after the end of the Eocene volcanism, heat flow faded to a lower level [4.12, 4.82], but there are now some local anomalies as in Landau/Pfalz [4.87]. The regionally high heat flow does not seem to be limited to the Rhinegraben. The Swabian volcanism and the second largest geothermal anomaly in Germany which has its maximum at Urach are most likely near surface events of one and the same thermal anomaly in the upper mantle.

Whereas regional high heat flow density can be explained as a deep anomalous heat source in the lower crust or in the upper mantle, local anomalies can be attributed to hydrothermal water migration and/or volcanic activity. The deep-seated faults of the Rhinegraben make it possible for thermal water to rise from several kilometers depth and fully explain local anomalies such as Landau/Pfalz [4.13, 4.85, 4.87]. The effect of large intrusive bodies on crustal rocks can be noted by observing the alteration of organic material in overlying rock. Such an example is found in the roof of the "Bramsche" intrusive in North Germany [4.14]. The intrusive body penetrated the crust about 100 million years ago. In the subsequent 2–3 million years it significantly increased the temperature gradient and with it the heat flow density in its vicinity.

Heat flow density is not perfectly constant in individual geotectonic units. Deviations from a single value within a unit can be considerable, as has been shown. A large granite body can significantly increase the heat flow density because of the high content of radioactive heat producers. Even so the various regions belong to a thermal tectonic unit defined as a heat flow province [4.64]. This can be demonstrated when dealing with inhomogeneities of the heat source distribution of the upper crust. One assumes that the heat flow density is constant from the lower crust (Q_0) below which no lateral inhomogeneities occur. It is possible that heat sources in these lower regions are either homogenously distributed or are negligibly small.

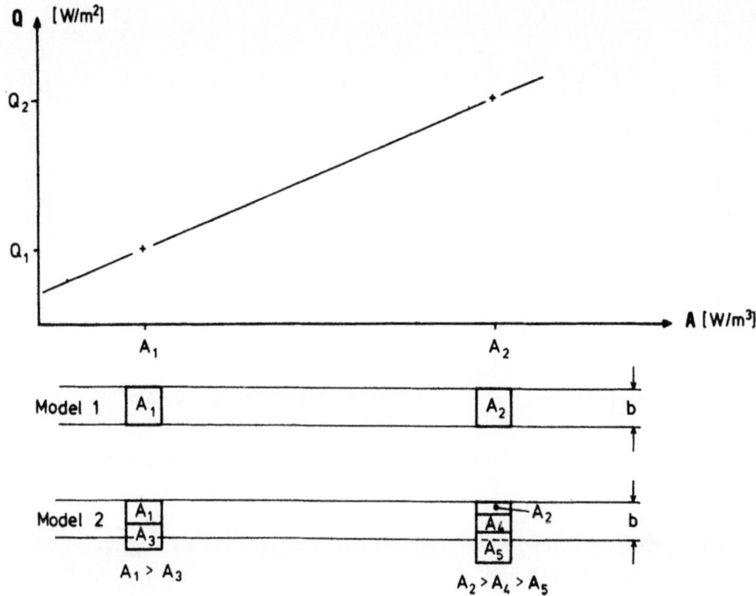

Fig. 4.15. Models for estimating the distribution of radiogenic heat sources of the crust from surface heat flow density after [4.64]

Within such a heat flow province there is a linear relationship between the surface heat flow density and the surface heat production A [4.64]:

$$Q = Q_0 + A\,b. \tag{4.23}$$

With the help of this relationship one can use two measurements of radiogenic heat production and heat flow density to estimate the thickness of the heat source distribution as well as the heat flow density from below (Fig. 4.15). With a greater number of measurements one can delineate the boundaries of the heat flow province.

The depth scale b implicitly assumes an identical distribution of uranium, thorium and potassium in the crust. It has been shown [4.43] that according to models of the geochemical evolution of the crust a different depth scale might be valid for each radioactive element. A simple model is proposed which explains the depth scale: the distribution of potassium has been fixed during the primary differentiation of the crust. Thorium gives the depth scale of magmatic or metamorphic fluid circulation, and uranium reflects late effects of alteration due to meteoric water [4.43].

Because the heat flow density generated by these heat sources is given as the integral of the distribution between the surface and depth z*

$$A(0)\,b = \int_0^{z^*} A(z)\,dz \tag{4.24}$$

different models of distribution A(z) can be considered as shown in Fig. 4.15.

In Europe there are more than 2000 heat flow density measurements [4.15, 4.40, 4.41]. The results of the largest part of these values [4.40] are shown by the frequency distribution in Fig. 4.16 from which the average value for Europe of $Q = 61$ mW/m^2 is calculated. The weak decline in the distribution curve towards higher heat flow

52

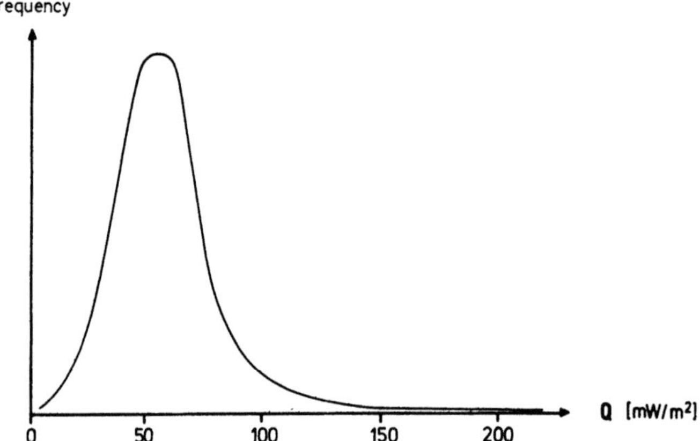

Fig. 4.16. Frequency distribution of measured heat flow density after [4.40]

density points to the relatively frequent occurrance of higher values which reach a maximum of 300 mW/m².

4.1.5.2 Secular Variation of Heat Flow Density

The oceanic heat flow density is greatest at a mid oceanic ridge, the site of creation of new oceanic crust, and decreases symmetrically to both sides. If effects of hydrothermal circulation in young ocean floor are avoided, the smooth heat flow density versus crustal age curve [4.73] shown in Fig. 4.17 is consistent with the model of sea floor spreading.

The most appropriate first approximation is the cooling model of an uniform half-space [4.20] which relates to the time t [Ma] as

$$Q = a\sqrt{b/t}$$

where a is a constant of about 500 mW/m² [4.51, 4.72] b a constant of 1 Ma and Q is the surface heat flow density. After 200 Ma the oceanic heat flow density seems to reach an equilibrium value of 38 mW/m² [4.75].

The continental heat flow density also correlates with geological age [4.16, 4.60, 4.75], even though the mechanism which causes the time dependence is totally different. With oceanic heat flow density, in addition to the part arising from the deeper earth, another contribution arises stemming from the cooling of the new oceanic crust. With continental heat flow density it is however the influence produced by radioactive heat producers. The continental heat flow density relates to the last orogenic event, the rate of erosion and the radiogenic heat distribution.

The regional heat flow density is lowest in Archean Europe with its Precambrian crust, somewhat higher in Mesozoic Europe, and has the greatest variation with sometimes very high values in Cainozoic Europe. This method of regional observations contains implicitly a time dependence on heat flow density first shown for data from Russia and Czechoslovakia as is illustrated in Fig. 4.18. The age of the geological unit is taken to be the age of the last geotectonic event.

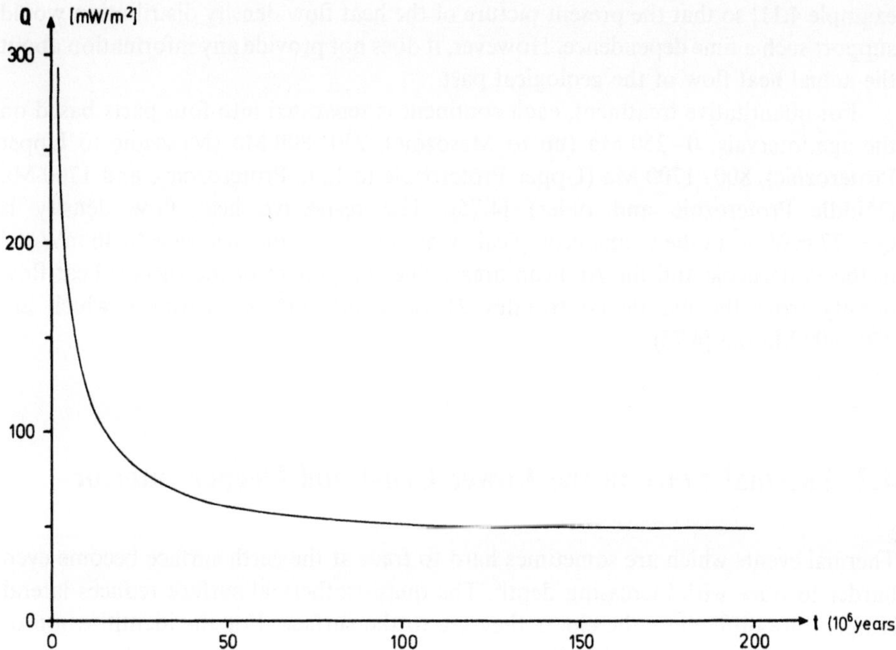

Fig. 4.17. Average heat flow density on the ocean floor and its dependence on the age of the crust after [4.73]

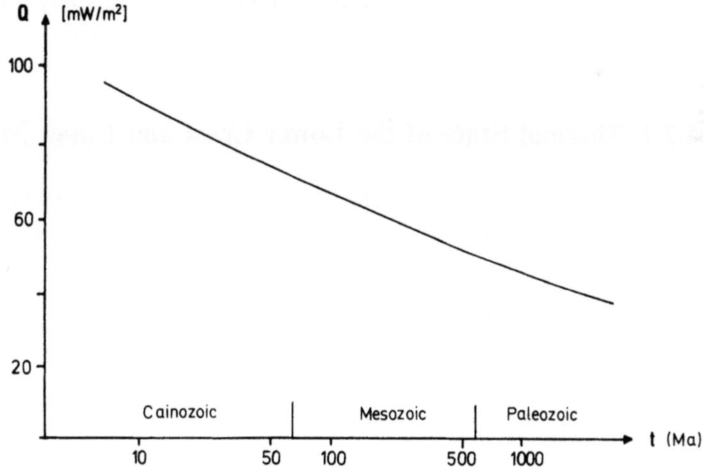

Fig. 4.18. Dependence of continental heat flow density of East Europe on the last geotectonic activity of a region after [4.16]

The tectonic event alone cannot produce heat which would significantly contribute to the regional increase of heat flow density. Neither can a cooling of the crust contribute to this heat because of the large time span in consideration. Instead heat supplied from the asthenosphere and/or the distribution of the radioactive heat sources most likely possesses this time dependency. The elements uranium, thorium and potassium seem to accumulate throughout earth history in the upper crust [for

54

example 4.11] so that the present picture of the heat flow density distribution would
support such a time dependence. However, it does not provide any information about
the actual heat flow of the geological past.

For quantitative treatment, each continent is separated into four parts based on
the age intervals: 0–250 Ma (up to Mesozoic), 250–800 Ma (Mesozoic to Upper
Proterozoic), 800–1700 Ma (Upper Proterozoic to Late Proterozoic), and 1700 Ma
(Middle Proterozoic and older) [4.75]. The respective heat flow density is
$Q = 77\ mW/m^2$ in the young geological areas, and this values decrease to $46\ mW/m^2$
in the Proterozoic and the Archean areas. The component of the surface heat flow
density from the mantle contributes 21 to $25\ mW/m^2$ in provinces which are
200–400 Ma old [4.75].

4.2 Thermal State in the Lower Crust and Deeper Interior

Thermal events which are sometimes hard to trace at the earth surface become even
harder to trace with increasing depth. The quasi-isothermal surface reduces lateral
temperature differences the closer they are to the surface. For the identification of
processes, which occur within the earth, it is necessary to consider indirect observations of seismology, gravimetry, geodesy and earth magnetism. Equilibrium conditions of phase changes are consigned to the earth's interior on the basis of anomalies
recognized by different geophysical methods. Only in this way is it possible to make
temperature estimates for the largest part of the inner earth.

4.2.1 Thermal State of the Lower Crust and Upper Mantle

Surface heat flow density (Q), which is measurable, contains essentially two parts after
corrections for interference factors from the near surface region. The first part stems
from the earth's interior and the second part from the decay of unstable isotopes of
the elements uranium, thorium and potassium. The latter decreases within the crust,
exponentially with depth. Different methods make it possible to estimate this part [for
example 4.10, 4.59] in order to calculate the heat flow density from the upper mantle.
A distribution of the heat-producing elements must be determined either from petrological models of crustal formation or from seismic as well as gravity models. In the
first case one starts at a surface value associated with the rock type (Sect. 2.3.1) and
assumes an exponential decrease of the radiogenic heat sources with depth (z). In the
second case one uses seismic velocity and density, both of which can be associated
with heat production as functions of depth (see Sect. 2.3.2), in order to obtain their
integrated distribution and to calculate the mantle heat flow density (Q_m)

$$Q_m = Q - \int_0^{\text{Moho}} A(z)\,dz. \tag{4.25}$$

Such a reduced heat flow density (Q_m) is a reference point for the temperature
distribution $T(z)$ below a certain depth z in the lower crust and in the uppermost

region of the earth mantle, assuming a value for thermal conductivity K

$$T(z) = T(z_0) + (z - z_0)\, Q_m/K. \tag{4.26}$$

In the following models a few temperature distributions will be given for different crustal types. The model calculations are based on assumptions for the distributions of thermal conductivity (compare Sect. 2.1) and radioactive heat production (compare Sect. 2.3). Presupposing a stable tectonic region one can use the stationary heat transfer equation (compare Sect. 3.1) to describe the temperature field. With that the temperature distribution can be calculated to a certain depth. This method is valid if, within meaningful geological time intervals, a stationary temperature field can be formed in the observed layer thickness. A two dimensional temperature field is calculated (Fig. 4.19 b) for an area between South Norway and Denmark after the model in Fig. 4.19 a using relaxation methods [4.4, 4.10]. The South Norwegian crust

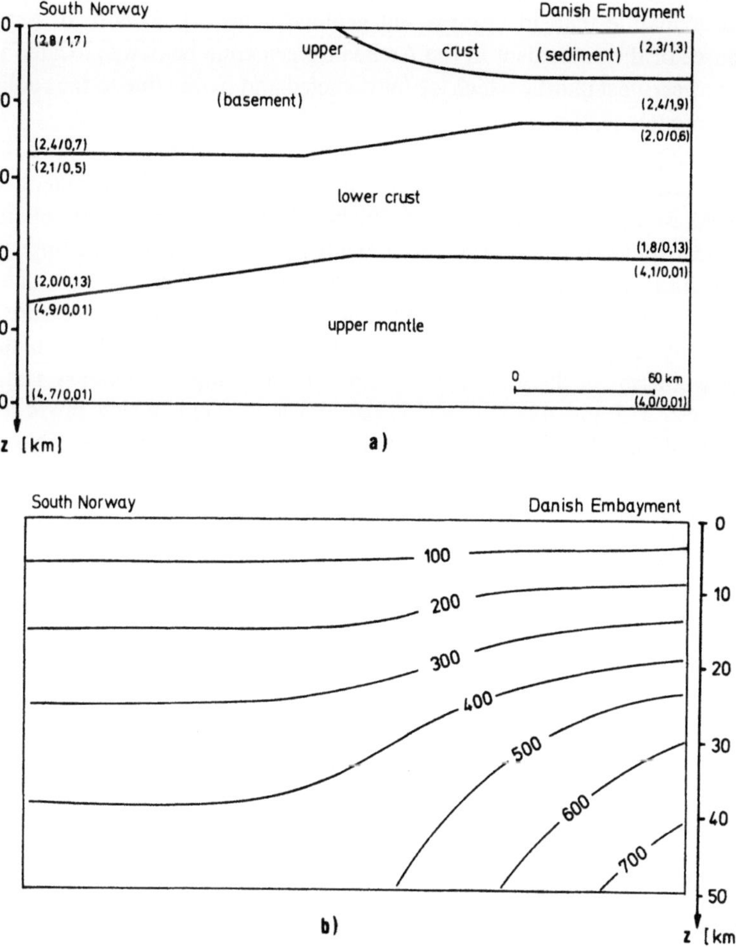

Fig. 4.19 a, b. Two-dimensional thermal model (Denmark) after [4.10]. a Upper lithosphere structure with thermal conductivity [W/m °K] and heat production in brackets [μW/m³]. b Resulting isotherms [°C]

56

with its upper mantle represents the craton, and Denmark represents normal continental crust. Depending on variations of thermal properties, the temperature distributions can vary considerably from each other [4.4]. The essential difference between the two crustal types, however, remains intact. In the old shields the melting temperature in the crust and upper mantle are never quite reached. In contrast, within the continental normal crust the melting temperature of siliceous rock within the lower crust can be attained (Fig. 4.20). As a rule by exceeding the melting temperature granite plutons are formed within the crust. These plutons often do not reach the earth surface but during their uprise cool off so much that they cool below the solidus temperature. With partial melting in the upper mantle a basaltic material is formed which frequently penetrates through deep crustal fault zones, thereby reaching the surface. This ascent from the depth of about 60 km in oceanic regions [4.23] occurs at such velocities, that basalt maintains nearly the same temperatures as its place of origin which is assumed to be about T = 1200 ± 100 °C [for example 4.77].

It is likely that, during the past, partial melting occured at shallower depths in the upper mantle than in more recent geological history. It is to be expected that the source of the volcanism in the Archean greenstone belts was located in a section of the uppermost mantle which is now depleted and frozen due to the cooling in the early Proterozoic [4.2].

The present day earth model is influenced by the knowledge concerning lithospheric plates and their drift over the asthenosphere. Lithospheric plates vary in thickness between about 50 and 200 km. For the thermal state of the lithosphere/ asthenosphere it is important to make the supposition that the upper surface of the transition zone is an isotherm (T = 1200 ± 100 °C). This is close to the melting point of the material of the lower lithosphere [4.17]. Such a temperature decreases the viscosity to such an extent that the drifting of plates is possible. Further indications for isotherms in the mantle are seismically determined boundary surfaces at depths between 375 and 425 km, at 550 km and at 670 km, which prove a high velocity gradient and therefore show transition zones of pressure phases. No changes in the

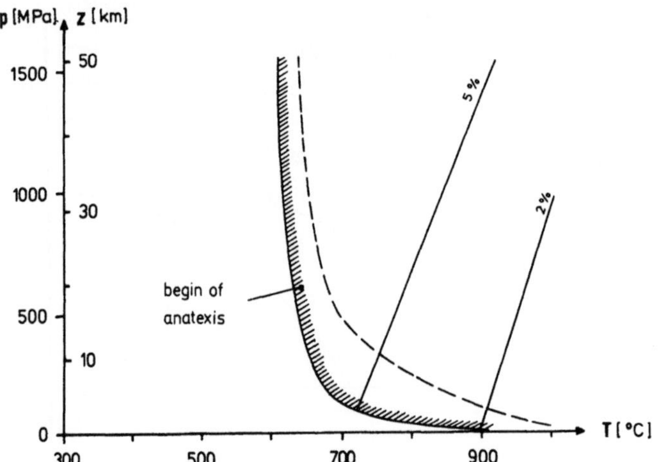

Fig. 4.20. Onset of melting in muscovite granite (———) at water saturation, 5% and 2% water content. Melting for water saturated gabbro is also shown (– –) [4.88]

chemical composition are required to exist. The region between the Mohorovičić discontinuity and the 400 km seismic discontinuity is designated as the upper mantle. The boundary surface between the upper and lower mantle is taken for a phase transformation [4.62, 4.63] in which the Mg-silicate $Mg_2 SiO_4$ changes from an olivine structure to a spinel type structure with a more dense packing. That is, the ratio of the actual volume of the lattice structure to the volume of the ions decreases. The pyroxene chain silicate converts to a new type of garnet structure [4.62]. This compression of material is connected with an increase in density of 8–10 %.

4.2.2 Thermal State of the Lower Mantle

The discontinuity surface at the boundary between the upper and lower mantle at about 400 km depth is a further fixed point for determining the temperature distribution of the mantle. The temperature at a pressure corresponding to about 400 km depth is estimated to be $T = 1300 \pm 150\,°C$ [4.77]. Because this boundary surface is at relatively constant depth ($375 \leq z \leq 425$ km), one can assume that at this depth there are no significant lateral temperature differences which can be traced back to distinct structure of oceanic and continental lithosphere.

In the region below 200 km depth there are no larger lateral temperature differences greater than about $200\,°C$, which in this case are due to mass convection in the asthenosphere [4.77].

The very distinct discontinuity of the P-wave velocity at about 670 km depth is compatible with the condition of a further material density increase, in which the spinel structures of $(Mg, Fe)_2 SiO_4$ converts to metal oxide structures with closest spherical packing and the high pressure modification of SiO_2 stishovite [4.29]. This phase change is given at a depth of about 670 km and an equilibrium temperature of $1600 \pm 400\,°C$. Further possibilities of phase changes lead to the same octahedral binding of silicon atoms with oxygen atoms in contrast to tetrahedral binding with lower pressure modifications [4.62].

Below 670 km depth seismically weak inhomogeneous regions of the lower mantle with further surface discontinuities could indicate phase changes as well as alterations in the chemical composition. At this time it is not possible to recognize points of references for fixed temperature points.

It is assumed that in the lower mantle there is a temperature gradient which corresponds to an adiabatic temperature distribution. Upwelling masses expand under lower pressure with decreasing density and cool off. Downwelling masses, on the other hand, become compressed and heat up. In such a system no changes take place with respect to the internal energy. The conditions for the radial adiabatic temperature gradient dT/dr is formulated in Thomson's equation:

$$dT/dr = -g\alpha T/c_p. \tag{4.27}$$

This equation is derived [for example 4.46] from the statistical basic equation in which the radial pressure gradient is proportional to the acceleration of gravity and density and from the First Law of Thermodynamics. In Thomson's equation g is the acceleration due to gravity, α the volume expansion coefficient, T the temperature and c_p the

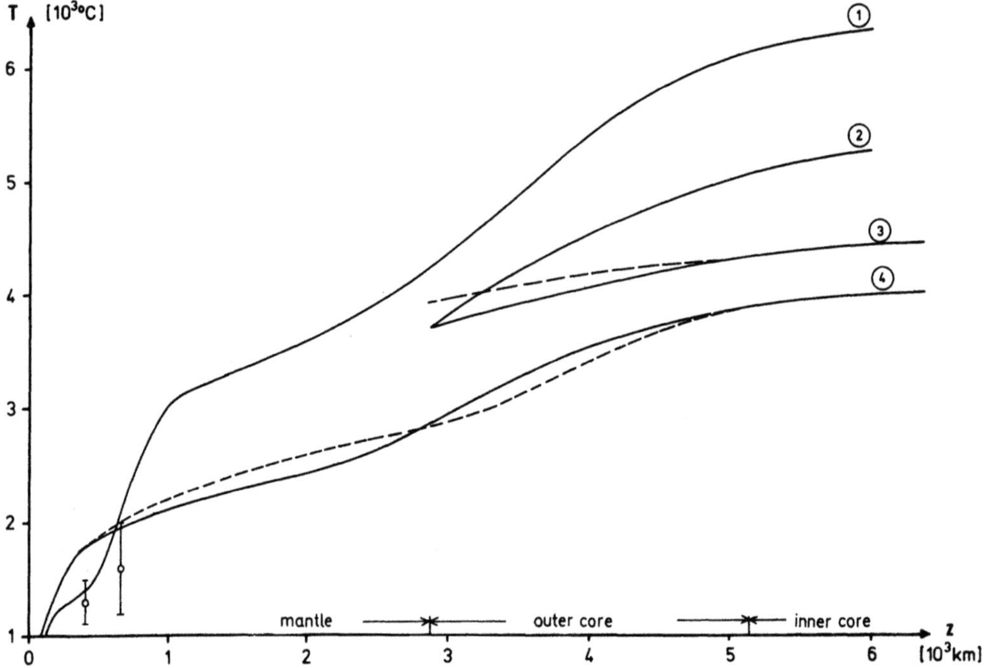

Fig. 4.21. Temperatures from phase transitions (■) and temperature distribution in the earth's interior based on

1) electric properties of the interior [4.83]
2) thermodynamic assumptions [4.61]
3) melting temperatures of iron (———) with adiabatic temperature distribution (– – –) [4.42]
4) physico-chemical and physical assumptions (———) with corresponding melting temperature (– – –) [4.79]

specific heat at constant pressure. The adiabatic temperature gradient amounts to

$$dT/dr \approx -0.5\,°K/km.$$

It is somewhat smaller than the isoviscous temperature gradient ($dT/dz \approx 0.7\,°K/km$) which is assumed to govern the thermal regime of the lower mantle. The expected conductive temperature gradient for the mantle is $dT/dz \approx 1.2\,°K/km$ [4.25].

A simple scaling analysis of the energy equation, coupled with considerations of a viscosity law supports the following mantle model: a lithospheric thermal boundary layer (basal $T \approx 1200\,°C$) a low viscosity asthenosphere, an isoviscous lower mantle, and a small basal thermal boundary layer with a temperature between $2600\,°C$ and $2900\,°C$ [4.25].

With the fixed temperature points in the upper mantle and an adiabatic temperature gradient in the lower mantle a temperature distribution results as shown in Fig. 4.21.

In the lowermost regions of the mantle, at depths of about 2550–2900 km the velocity gradient of the longitudinal waves are very low, which means a high temperature gradient when dealing with homogeneous material [4.1].

Calculated temperature distributions deviate from one another depending on the assumptions of chemical compositions and physical properties.

4.2.3 Thermal State of the Core

The center of the earth can be divided on the basis of seismic data into a solid nucleus ($5150 \leq z \leq 6371$ km) which probably consists of an iron-nickel alloy and a liquid outer core. The outer core may consist of iron and sulfur. The surplus, probably iron, crystallizes out and maintains a quasi-equilibrium with the melting process. The changes are only recognizable in geological time and possibly are responsible for the reversal of the magnetic field of the earth. The exact characterization of the core material is difficult because of the imprecise determination of its density distribution and its seismic parameters. If one assumes pure iron for estimating the melting temperature of the core [4.8, 4.35, 4.42], the melting point at the core-mantle boundary is calculated at $T = 4800\,°C$. On the other hand, an eutectic mixture of iron and sulfur [4.79] reaches a melting point of 2600 °C at the same boundary surface. The average value of these two results is $T_M = 3700\,°C$. In the fluid outer core actual temperatures exceed the melting temperature. In the solid inner core these melting temperatures slightly exceed actual temperatures. A few temperature distributions are shown in Fig. 4.21. They are based on calculations:

1. from the distribution of electric conductivity in the mantle [4.83] with a core temperature as in [4.28],
2. the thermal equation of state with assumptions for chemical composition and temperature and pressure dependent density changes
3. the melting temperatures of iron [4.42] and
4. the thermodynamic supposition of melting temperatures for an eutectic iron-sulfur mixture and pressure caused changes in the electron shells of potassium as a heat source [4.79].

The calculations are interpreted in terms of physical properties which are extrapolated from laboratory experiments to pressures and temperatures presumed to exist in the core. The present knowledge of thermal conditions within the core could be revised substantially through better laboratory experimentation and improved geophysical measuring techniques.

At the boundary between the inner and outer core it is very probable that the melting temperature of core material is equal to the adiabatic distribution. The gradient of the melting temperature of the outer core must therefore be greater than the adiabatic temperature gradient, in order to fulfill the condition that the outer core is fluid. However, the core material is certainly not a chemically pure substance. Therefore there is no definite melting point, but rather a melting zone in which the fluid phase remains in equilibrium with a solid phase. The liquid phase of the outer core may contain up to 30 % suspended solid particles, without influencing the seismic parameter [4.1].

The low temperature gradient concluded from the melting point calculations suggests an almost isothermal inner core. However, it also suggests such a small temperature change in the outer core, that large scale convection which produces the magnetic field as a dynamo effect could not occur.

A possible explanation for maintaining convection without changing the perception of the inner core and lowermost mantle would result from heat-producing events or primary heat sources in the outer core and heat-consuming events in the lowermost mantle. Iron could crystallize out [4.79] at the border of the outer/inner core and thereby release heat of crystallization. Also potassium could be present as a radiogenic heat source in the fluid phase of the outer core. Heat sinks at the lower mantle boundary could possibly be the heat of solution, which would be exhausted if the possibly iron-rich lower mantle would be converted in part to the liquid phase of the outer core. The change in the chemical balance could lead to crystallizing out of iron, thus causing the inner core to grow at the expense of the mantle.

4.3 Thermal Aspects of Plate Tectonics

The new global tectonics developed in the last decades stems from Wegener's theory of continental drift. Initial hypotheses about the oceanic lithosphere have led to the present understanding of plate tectonics [e.g. 4.21, 4.34, 4.37, 4.49].

Plate tectonics requires division of the earth different from the separation into a crust and mantle in the outer regions. The material break-down, which allows a marked division at the Mohorovičić discontinuity, is substituted with a geodynamic one, which combines the crust and part of the upper mantle into the lithosphere. The boundary between the lithosphere and the asthenosphere is a zone of low viscosity. This boundary uncouples to a large extent the mechanical processes of lithosphere from those occurring below.

The lithosphere consists of individual bounded plates, which do not rest rigidly on the underlying asthenosphere, but move relative to one another. The movement of the oceanic lithosphere with its geological and geophysical events is particularly instructive. Here along the mid oceanic ridges partially melted mantle material rises to the surface. Oceanic plates drift away on both sides of the ridge, submerge along the continental boundaries under the light sialic crust of the continents and again become assimilated into the upper mantle. In contrast to the above-mentioned ocean-continent boundary, the boundary of two continental lithospheres is the site of horizontal (strike-slip) motion or of convergence with partial overthrusting or also subduction.

The visible surface signs of plate movement do not easily reveal the driving force from the interior, which are constant neither in time nor place. Magnitude and direction of plate movements can be subject to episodic events.

There is, however, no doubt that the driving mechanism of plate movement is essentially of thermal nature supported for example by gravity forces, which occur with subduction. These forces alone do not fully explain the tectonic connection through time between mountain building on the continental lithosphere and the spreading of the oceanic lithosphere [e.g. 4.57].

Following the discovery that the heat producing unstable isotopes of uranium, thorium and potassium occur in all rocks, it is considered likely [4.36] that convection currents occur in the earth mantle, maintained by these radioactive heat sources. The thermally induced and large-scale convection cells, which have a radius of several thousand kilometers, require much too great circulation times. Another argument against a model with such large convection cells is the uneven movements of the lithosphere plates which would have a large time constant. It seems more likely to be thermal energy in the asthenosphere, for example, activated by differentiation processes in the broadest sense, which causes a mass convection in connection with a thermal event. Numerous signs of mass movements (mid oceanic ridge, mantle diapirs, hotspots) confirm that the more or less radially directed convective mass transport is of singular nature and that the explanation is found in the thermal development of the planet.

The radial upwellings of partially molten mantle material to the surface differ strongly in their intensities, being strongest at the mid oceanic ridge and only weak where it is traced under the continental lithosphre. The latter is shown by an elevated "Moho" as for example in the upper Rhinegraben. The convective upwelling is always accompanied by a positive thermal anomaly. The convective downflow in the subduction zones should be accompanied by a corresponding negative thermal anomaly, which, however, is totally dominated by the tectonic and magmatic side effects.

The physical principle which adequately explains the release of thermal energy in the asthenosphere has not yet been found. Many model observations for the mechanism of plate movement are based on simplified ideas of the classical convection theory. The thermal events on the surface and their connection with the movement and thickness of the lithosphere plates may be better understood through improved models incorporating

1. the temperature dependence of the viscosity,
2. horizontal temperature gradients,
3. radiogenic heat sources,
4. frictional heat of shearing and
5. phase changes [e.g. 4.17, 4.56, 4.69, 4.74].

Most of the well-known geothermal anomalies are located at or near lithospheric plate boundaries [4.65]. For oceanic lithosphere the simple plate models of the creation of lithospheric plates at mid oceanic ridges and their slow cooling process agree with the measured heat flow density. Oceanic heat flow density Q is inversely proportional to the square root of the age t of the corresponding lithosphere

$$Q \sim \sqrt{1/t}. \tag{4.28}$$

This simple model also explains the connection between the topography of the lithosphere, that is, its depth D below sea level and its age t

$$D \sim \sqrt{t} \tag{4.29}$$

[4.19. 4.55]. Systematic deviations of the measurement values from model results, which cannot be traced back to intensive water circulation [4.19], show up only after t > 70 million years. For the older parts of plates the measured elevation values are higher than expected. Increasing shear resistance can become significant in regions distant from the ridge as manifested through increasing frictional heat. After such a

62

correction the validity of the above named models is expanded to a greater time (t > 70 million years) [4.69].

The heat flow density distribution in an area of a subducting plate shows a distinct pattern (Fig. 4.22): the values are low, i.e. 20 to 40 mW/m² between the trench and the aseismic front. A transition zone occurs between the aseismic front and the volcanic front. High heat flow densities, i.e. 60 to 120 mW/m² prevail in the area behind the volcanic front [4.38]. Besides the conductive heat, frictional heating along the surface of the downgoing slab contributes to the thermal regime of subduction. The subducting oceanic lithosphere contains water as pore fluid in the sedimentary cover a well as in minerals such as amphibole and others. During subduction the water is gradually released and supports the magma generation by partial melting. The magma ascends giving rise to the volcanic front above the downgoing plate (Fig. 4.22). The subducted material is cooler than the adjacent asthenosphere which causes a depression of the isotherms and a remarkable horizontal component of the heat flow.

Another possibility for explaining the differences between the theoretical values of the simple cooling model and the measured values for the lithosphere age t > 70 million years is shown in a further model [4.56]. It is not based on a simple cooling half space model, but is based on plate model with boundary conditions that the surface heat flow density and the depth of the ocean floor reach a constant boundary value. According to this model the heat flow density decreases exponentially with age and reaches an asymptotic boundary value. For times up to t = 70 million years both above named models satisfy the observations. However, the modified plate model is in agreement with observations for all times. From the last model, calculations yield an isostatic compensated lithosphere thickness of

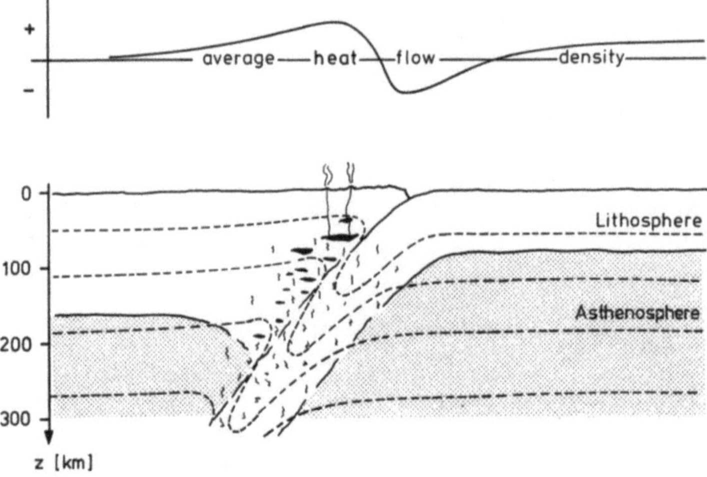

Fig. 4.22. Thermal structure at a plate boundary, where an oceanic plate is sliding under a continental plate, causing an anomalous heat flow density at the surface. The negative part is often masked due to hydrothermal activity. The curvature of isotherms is qualitatively shown as dashed lines

125 $\pm$ 10 km and a boundary surface temperature of T = 1350 $\pm$ 275 °C at the transition to the asthenosphere.

In a thermal model for the continental lithosphere one can assume a heat transfer from the asthenosphere to the lithosphere with a value Q [4.17]. This heat flow density is considered globally constant under both the oceanic and continental lithosphere. The boundary between the lithosphere and the asthenosphere is an isotherm surface, which is the result of a thermodynamic equilibrium. Each change in the thermal condition of the asthenosphere requires a shifting of the isotherms and thus also the boundary between the lithosphere and asthenosphere. The thickness of the lithosphere according to this model is caused by the thermal condition of the asthenosphere. If the heat from the asthenosphere increases, the lithosphere becomes thinner and a new equilibrium is achieved. After the anomaly fades, that is the heat sources vanish or the lithosphere plates drift from the anomaly, then the lithosphere cools off and the thickness increases. This model agrees with the heat flow density in North America [4.17].

The thermal models of the lithosphere consider plates as passive portions whose movement is forced through mass convection within the asthenosphere. With hotspots (or radial upwelling of partially molten asthenosphere material) such forces can be assumed [4.52], which are connected with the origin of plate movement. Most hotspots lie in the vicinity of mid oceanic ridges and that there are always hotspots situated at the boundary of three large plates. The upward directed asthenosphere flow transports the plates away from their place of origin. There are signs [4.52] that the asthenosphere becomes active before the continents are divided by the upwelling. A thermal anomaly involves much more than volcanism which is simply a surface expression of the phenomena. Locations where partially molten asthenosphere material wells up are also indicated by gravity data and regional topography around the place of origin. The gravimetric anomaly and the topography are possibly a measure for the amplitude of the thermal anomaly in the asthenosphere [4.52].

The upwellings are apparently not influenced by the lithosphere. They leave their visible marks of past activities on the continents as well as the oceans: the ringdike complex in SW Africa, the flood basalts of the Parana Basin in S. America, the Dekkantrapp basalts in India, the island of Hawaii. The asthenosphere upwelling can be intensive enough to divide a lithosphere plate and thus to start the formation of a new ocean as is shown along the Red Sea and the Gulf of Aden [4.58].

The upwellings from the asthenosphere appear to be fixed in space. In some places their volcanic manifestations penetrate the lithosphere, so that movement of the lithospheric plates can be traced in time. The relationship between upwelling corresponding to the hotspots and downwelling currents is not yet known. Asthenospheric upwelling in a larger sense is an expression of the thermal evolution of the earth. It contributes on one hand to a convective cooling of the asthenosphere and on the other hand to a material differentiation of the earth's mantle.

Supplementary Problems

4.1 A conical mountain has a height of 1000 m and a diameter of 3000 m at the base, which is taken as datum plane (see Fig. 4.4). Determine the correction of measured temperature gradient (d T/d z = 25 °C/km) at a depth of 2000 m beneath the top of the mountain.

4.2 Thermal water (T = 80 °C) ascends along a vertical fault from a depth of 6000 m. Determine its thermal effect 200 m beneath the surface and at a distance of 2000 m from the fault.

4.3 Determine the lateral temperature difference between the temperatures in a layered sequence of sedimentary rocks and a homogeneous sandstone as shown in Fig. 4.10. The conductivity of sandstone is $K_s = 3.2$ W/m °K. The conductivity and thickness of each layer are: $K_1 = K_s$, $h_1 = 200$ m; $K_2 = 2.5$ W/m °K, $h_2 = 200$ m; $K_3 = 2.0$ W/m °K, $h_3 = 1500$ m; $K_4 = 0.26$ W/m °K, $h_4 = 10$ m. The conductivity of the substratum is K_s. Assume a heat flow density of 80 mW/m².

5 Methods for Determining Temperature

A knowledge of temperatures within the earth is essential for the proper under-standing of many geologic, mineral-petrographic and geophysical properties and processes. For example, temperatures govern the onset of melting, fix equilibrium conditions of certain mineral paragenesis and anomalous temperatures define heat flow anomalies.

However determining temperatures pertinent to geologic processes is not simple. First, most of earth's interior is not accessible for direct measurement. Even the near surface layers, down to a depth of a few kilometers, require expensive drilling before direct temperatures can be measured. For this reason, indirect methods must also be used for temperature estimations. Furthermore, one may not be interested in the present thermal state of particular crustal regions but rather that of the distant past. Such is the case for example, for questions dealing with the metamorphism of rocks or formation temperatures of certain mineral paragenesis. Fortunately the tempera-ture dependence of several chemical reactions in the broadest sense or the temperature dependence of certain physical quantities can be used to construct a geothermometer obtaining a picture of space and time temperature behaviour of the earth's interior. This chapter is devoted to the description of several geothermometers and their use in geology and geophysics.

5.1 Geothermometers for Evaluating Reaction Temperatures

Physical-chemical equilibrium conditions can theoretically be used to calculate a particular equilibrium temperature. However, many phases are in equilibrium over a broad temperature interval. The presence of a specific mineral paragenesis, that is an assemblage of minerals originating in the same geochemical environment, therefore provides little precise information about their formation temperatures.

The inclusion of trace elements in the crystal lattice of different minerals or the relationships of specific ions and isotopes in a phase allows a more sensitive deter-mination of the formation temperatures. Also the solubility of solid materials such as quartz and minerals containing Na, K and Ca permit a retrospective calculation of solution temperatures. Finally, organic inclusions in sedimentary rocks are very sensi-tive to temperatures, changing the chemical constitution irreversibly.

5.1.1 Solution Equilibria as Temperature Indicators

In a saturated porous rock a solution equilibrium established after a certain time and at a given temperature. Only a very small part of the rock component enters into

66

solution. Individual minerals and amorphous components such as glass phases in eruptive rocks or opal have very different solubility. Salts are easily soluble, whereas rock-forming minerals such as quartz and feldspar are much less soluble. Ignoring the fact that various dissolved components influence each others' solubility, for each component the integrated form of Van't Hoff's equation is valid. That is

$$\ln C = \Delta H/(R\,T) + x_1 \tag{5.1}$$

where C is the equilibrium constant, ΔH is the enthalpy, R is the universal gas constant, T is the absolute temperature and x_2 is an integration constant. It is evident that the logarithm of the equilibrium constant of the dissolved substance is inversely proportional to temperature.

5.1.1.1 The SiO$_2$-Thermometer

The SiO$_2$ content of thermal water is frequently used to determine the temperature prevailing in subsurface water reservoirs. Even though there are many limiting factors for the validity of equilibrium conditions, the SiO$_2$-thermometer yields a useful temperature estimation. However, temperatures calculated using the SiO$_2$-thermometer are generally too low [5.41]. Deviations lie in part with the analytical determination of silicic acid (H_2SiO_3) and also in the following assumed, but not always satisfied, conditions [5.24]:

(1) The temperature-dependent reaction (i.e. solution equilibrium for quartz and other SiO$_2$ forms) takes place only at reservoir depth and no further reaction occurs.

(2) Water production of a thermal spring is at least 200 l/min, because otherwise SiO$_2$ would have time to precipitate out at lower temperatures.

(3) Mixing of thermal waters with water of other layers does not take place as it could lead to a change in concentration.

Further inaccuracies arise because of varying solubilities of different SiO$_2$ polymorphs such as quartz, chalcedony, christobalite and amorphous SiO$_2$ [5.21, 5.22]. Also the pH value of water plays an important role. The solubility of amorphous SiO$_2$ (opal) is considerably greater than that of the fine-grained and porous chalcedony, whose solubility is again larger than that of quartz. Temperature estimations using the SiO$_2$ geothermometer can only be precise, if the starting material is known. Experience shows that low temperatures (to $T \approx 100\ °C$) should be calculated with the chalcedony solubility and higher temperatures with that of quartz [5.41]. According to equation (5.1), one calculates temperatures of the solution equilibrium [5.41, 5.70] for

$$\text{Chalcedony:} \quad T[°C] = \frac{1032}{4.69 - \log c} - 273 \tag{5.2}$$

$$\text{Quartz:} \quad T[°C] = \frac{1315}{5.205 - \log c} - 273. \tag{5.3}$$

The SiO$_2$ content c is given in [mg/l]. Both functions are illustrated in Fig. 5.1. The given equations can be used up to 250 °C. Above this temperature the accuracy is rather weak [5.21].

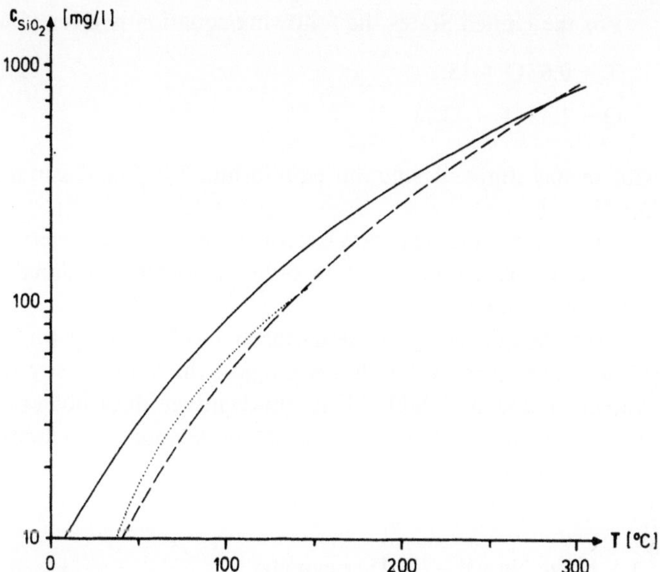

Fig. 5.1. Solubility of SiO_2 in water
——— chalcedony as in [5.41, 5.70]
– – – quartz
····· calibration curve for SiO_2 content of thermal water in Germany after [5.29]

For the Federal Republic of Germany SiO_2 content is nearly always below 100 mg/l. It has been shown that the calculation of the equilibrium temperature, based on quartz solubility, can be approximated by the following equation [5.29]:

$$T[°C] = 10.28\,(c)^{0.5607}. \tag{5.4}$$

The SiO_2 content (c) is again given in [mg/l]. This curve is also shown in Fig. 5.1.

Under stationary conditions of heat conduction (Sects. 1.2 and 1.3) one can calculate the depth at which water reaches the solution equilibrium:

since $\quad T = z\ dT/dz + T_0$

and $\quad Q = K\ dT/dz$

the depth (z) is given by:

$$z = (T - T_0)\ K/Q. \tag{5.5}$$

T is the calculated temperature from SiO_2 content, T the mean annual temperature of the surface, K the thermal conductivity and Q the heat flow density. However, the error caused by uncertainties in average thermal conductivity is even greater than that of determining temperature.

If in equation (5.5) the depth of water circulation, i.e. reservoir depth (z) and thermal conductivity (K) is taken to be constant, then there is a linear relationship between the silica thermometer temperature T and heat flow density Q. In areas without any heat flow data this can be of considerable help in interpolating data.

68

For the United States the following equation is given [5.64]:

$$T = 0.67\,Q + 13.2 \tag{5.6}$$

$$Q = 1.49\,(T - 13.2) \tag{5.7}$$

with temperatures of solution equilibrium $T[°C]$ and surface heat flow density of $Q[mW/m^2]$.

On account of large possible errors in temperature determination, a large number of single values are necessary in order to obtain a regional heat flow density with sufficient accuracy.

Even though the SiO_2 thermometer has been used for some time with success [5.22], it has considerable disadvantages which are mostly releated to the absolute content of dissolved SiO_2. This disadvantage does not exist for the $Na-K-Ca$ thermometer described in the next section because the contents are only used as ratios in calculations.

5.1.1.2 The Na−K−Ca Thermometer

A geochemical thermometer based on equation (5.1) was developed [5.21, 5.23] which uses the sodium, potassium and calcium concentrations of thermal water in calculating the temperature of the solution equilibrium. By using the ratio of their concentrations the error margin is small even though for water containing Na, K, and Ca, the same limitations [5.24] apply which have been given in the previous section for the SiO_2-thermometer. The equilibrium temperature of the solution is given [5.41] by:

$$T[°C] = \frac{1647}{2.24 + F(T)} - 273 \tag{5.8}$$

with

$$F(T) = \log(c_{Na}/c_{K}) + x\,\log(\sqrt{c_{Ca}}/c_{Na}).$$

x amounts to ⅓ if the mol-ratio $\sqrt{c_{Ca}}/c_{Na} < 1$ and x amounts to ⁴⁄₃ if $\sqrt{c_{Ca}}/c_{Na} > 1$. If in the latter case the calculated temperature $T > 100$ °C, then x should be set at ⅓ in order to obtain an equilibrium temperature. The function $F(T)$ dependent on temperature (T) is shown in Fig. 5.2.

The temperature obtained from the Na-, K- and Ca-content of groundwater is often higher than that which is determined by SiO_2 content [5.41]. There may be a number of reasons for this. For one, solubility of each mineral considered is quite variable. Furthermore, SiO_2 can precipitate out during mixing of cold water with thermal water. Finally, requisite minimum discharge of 200 l/min is a large requirement for a single spring. But even the $Na-K-Ca$ thermometer has regular errors. If the thermal source is within a sedimentary basin, the salinity of porefilling fluid is important. Also considerable perturbations of the thermometer can be caused by monomineralic salt deposits.

A map has been produced for the Federal Republic of Germany showing $Na-K-Ca$ temperatures [5.74]. Large scale structures where $Na-K-Ca$ temperatures rise considerably above their surrounding values include the entire Upper Rhi-

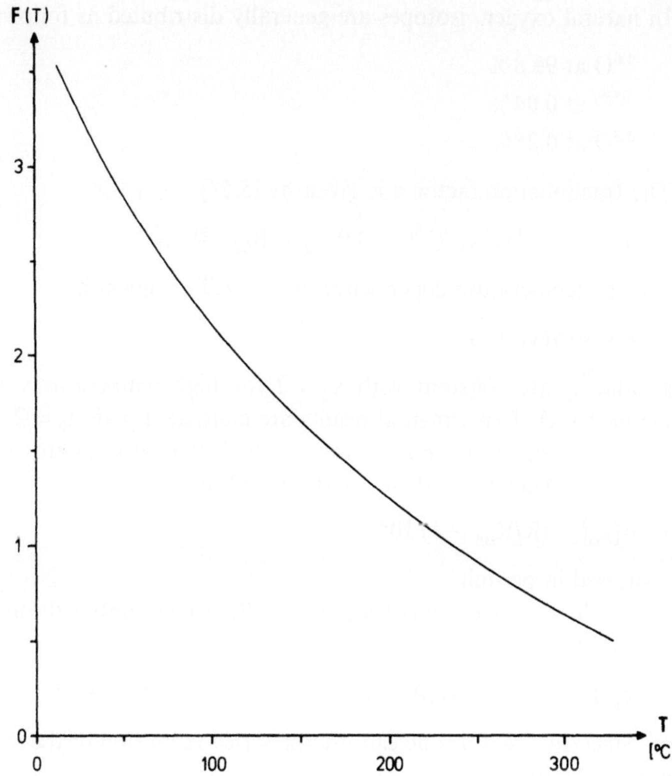

Fig. 5.2. Calibration curve for application of Na − K − Ca thermometer according to equation (5.8) after [5.41]

negraben (T = 150 °C) as well as the Swabian and Franken Jura (T = 100 °C). Because the calculated temperatures cannot give clues to the depth in which the solution equilibrium took place, one cannot conclude on the basis of these data alone that certain thermal sources are to be found within a geothermal anomaly. Only with the help of heat flow density and thermal conductivity measurements (see Sect. 5.1.1.1) can one estimate equilibrium depth.

5.1.2 Isotope Ratios as a Geothermometer

Fractionation of stable isotopes of various elements in solutions, minerals and rocks has been recognized of late as an important geothermometer because of the temperature dependence of the fractionation factor [5.33, 5.53]. It is mainly the isotope ratios (R) of the elements hydrogen (D/H), carbon (^{13}C/^{12}C), oxygen (^{18}O/^{16}O) and sulfur (^{34}S/^{32}S) which are used for temperature determinations. The isotope ratio is determined by mass spectrometer analysis. One must assume that the analyzed components reflect the equilibrium condition of an isotope exchange reaction, as for example that of the oxygen exchange reaction:

$$H_2^{18}O + \tfrac{1}{2}C^{16}O_2 \rightleftharpoons H_2^{16}O + \tfrac{1}{2}C^{18}O_2.$$

70

In natural oxygen, isotopes are generally distributed as follows:

^{16}O at 99.8 %
^{17}O at 0.04 %
^{18}O at 0.2 %.

The fractionation factor α is given by [5.53]:

$$\alpha = (^{18}O/^{16}O)_{CO_2}/(^{18}O/^{16}O)_{H_2O} = R_{CO_2}/R_{H_2O} \tag{5.9}$$

and the temperature dependence of $\alpha = \alpha(T)$ is given by

$$\alpha \sim \exp(x_1/T^{x_2}). \tag{5.10}$$

x_1 and x_2 are constant with $x_2 = 2$ for high temperatures and $x_2 = 1$ for lower temperatures. Experimental results are consistent with $x_2 = 2$.

The isotope ratio (R) is compared with that of standardized samples (R_{std}). The difference, known as "delta" and defined as

$$\delta[\text{‰}] = (R/R_{std} - 1)\,10^3 \tag{5.11}$$

expressed in per-mil.

The logarithmic form of equation (5.9) can be combined with equations (5.10) and (5.11) yielding:

$$x_1/T^{x_2} \sim \ln\alpha = \ln(10^{-3}\,\delta_{CO_2} + 1) - \ln(10^{-3}\,\delta_{H_2O} + 1).$$

Since $10^{-3}\,\delta \ll 1$, one can use the series expansion of the logarithm and truncate it after the first term:

$$10^3\,x_1/T^{x_2} \sim 10^3\,\ln\alpha + x_3 = \delta_{CO_2} - \delta_{H_2O} \tag{5.12}$$

x_i $(i = 1, 2, 3)$ are constants. The difference between the relative isotope ratios is inversely proportional to temperature (T^{x_2}), as is seen from equation (5.12).

For calculating an equilibrium temperature (T) in the $CaCO_3 - H_2O$-system for calcium-containing shells in marine organisms, the determining equation [5.19] is:

$$T[°C] = 16.5 - 4.3\,(\delta_c - \delta_w) + 0.14\,(\delta_c - \delta_w)^2. \tag{5.13}$$

The relative oxygen-isotope ratio in CO_2, which stems partly from $CaCO_3$ (δ_c) and partly from water (δ_w) corresponds to the equilibrium of the isotope exchange reaction of the $CO_2 - H_2O$-system. The sensitivity of the difference of $\delta\,^{18}O$-values amounts to about 0.2 per-mil/°C. The application of this method is limited to ocean temperatures.

Definitive equations do not exist for the calculation of formation temperatures of crustal rocks. Nonetheless isotope studies provide some information about their genesis [5.53]. For example, plutonic granites have higher $\delta\,^{18}O$-values than their equivalent extrusive [5.65]. The reasons may be that intrusive rocks are derived from ^{18}O-richer materials such as sedimentary rocks and/or a magmatic differentiation took place at a lower temperature. The $\delta\,^{18}O$ analysis of metamorphic rocks show that serpentinization of olivine can take place within a broad temperature interval beginning with the temperature at the earth's surface. Equilibrium temperatures of the isotope exchange reaction [5.53] indicates that chrysotile (fiberous serpentine) is formed at low temperatures. If water absorption takes place at higher temperatures

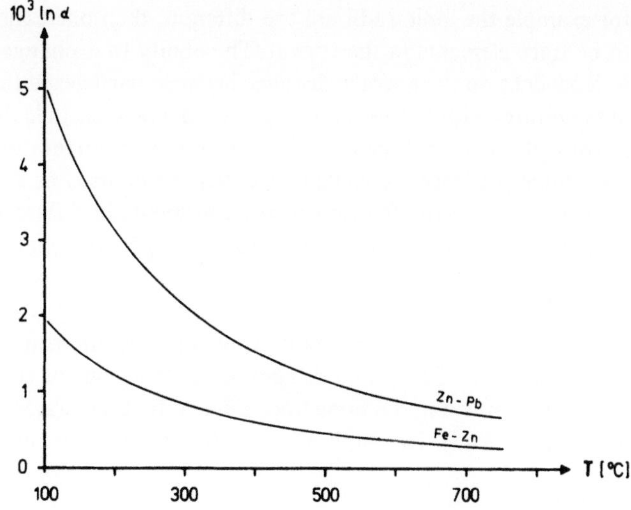

Fig. 5.3. Dependence of the fractionation factor α on temperature for the sulphide pair after [5.50]:
1. Zn − Pb sphalerite/galena
2. Fe − Zn pyrite/sphalerite

[200–400 °C], then antigorite (layered serpentine) forms. Epizonal metamorphic rocks are in general richer in ^{18}O than those of their mesozone and catazone counterparts and thus reflect a connection in which the ^{18}O content behaves inversely proportional to temperature.

The isotope ratio of sulfur is also used as a geothermometer [5.50]. Of the sulfur isotopes, ^{32}S is the most commonly occurring (about 95 % of natural sulfur) and ^{34}S (about 4 %) is the next most common. The ratio of both isotopes ($^{34}S/^{32}S$) is dependent of many factors. Apart from the pH-value and the oxygen – partial pressure the biogene factors are important in the low temperature regions. The influences of these biogene factors far outweigh the temperature dependence of the isotope exchange reaction. In spite of this an equilibrium temperature of the isotope exchange reaction has been obtained successfully for some sulphide ore minerals [5.50]. Fig. 5.3 illustrates the temperature dependence of the fractionation factor for two sulphide pairs (sphalerite/galena, pyrite/sphalerite). At the basis of this is the assumption that the ores were formed simultaneously from a hydrothermal solution which was in a state of equilibrium. The temperature then corresponds with the formation temperature of the ore in analogy to equation (5.13).

5.1.3 Trace Elements in Salts and Ores

When minerals crystallize out of a fluid phase, impurity ions can be built into the crystal lattice. The replacement of lattice components with impurity ions is called substitution. The smaller the difference in the crystal chemistry properties of two of these exchanging components, the greater the exchange affinity. This is most evident with solid solutions. In olivine, which is made up of the iron silicate fayalite (Fe_2SiO_4) and the magnesium silicate forsterite (Mg_2SiO_4), any number of magnesium ions may be replaced with iron ions. In contrast, the absorption capacity of bromine ions in a rock salt lattice is much smaller. If only a limited substitution can take place because for example the ionic radii are too different, then one considers the substituted ions to be trace elements in the crystal. The ability to exchange particles of varying sizes is dependent on the average distance between particles in the lattice, so that at higher temperatures with corresponding greater average particle distance, the concentration of trace elements may be greater than for lower temperatures. This property permits the content of trace elements in a crystal to be used as a suitable geothermometer.

If m_1 is the mole fraction of trace elements in a fluid phase and m_2 that in the crystal, then within a specific interval m_2 is proportional to m_1

$$m_2 = C m_1. \tag{5.14}$$

C is the proportionality constant in the linear relationship and is called the distribution coefficient. If a second mineral crystallizes out of the same fluid phase, at the same time absorbing the same trace element in its crystal lattice, one can eliminate m_1 in equation (5.14). For both minerals the following is then true

$$m_{21}/C_1 = m_{22}/C_2.$$

Using equation (5.1), which describes the temperature dependence of a distribution coefficient, one obtains for very small trace element contents and with constants x_i $(i = 1, \ldots, 6)$:

$$\ln m_{21} + x_1/T + x_2 = \ln m_{22} + x_3/T + x_4. \tag{5.15}$$

This leads to [5.71]:

$$\ln(C_1/C_2) = \ln(m_{21}/m_{22}) = x_5/T + x_6. \tag{5.16}$$

Equation (5.16) gives the temperature dependence of the trace element concentration. It is assumed that the concentration of the trace element considered in each mineral is limited and that at constant temperature an equilibrium has been maintained in the systems: (mineral 1)-(fluid phase) and (mineral 2)-(fluid phase).

Example of such a geothermometer are the bromine content in rocksalt (NaCl) and sylvite (KCl) for low temperature or the manganese and the cadmium content in the ore wurtzite (ZnS) and galena (PbS) for higher formation temperatures.

When salt water basins are cut off, the concentration of salt increases due to evaporation until finally the saturation point is reached and salt precipitates out. For sulfur-free water, rocksalt (NaCl) forms first. At higher brine concentration sylvite (KCl) forms. The onset of sylvite precipitation and correspondingly the solubility of sylvite are strongly temperature-dependent. At 0 °C crystal formation begins with concentration 55 times greater than ocean water and at 25 °C crystallization begins with concentration 64 times greater [5.7]. If bromine ions are present in the water, then they are built into the crystal lattices of rock salt and sylvite according to their respective substitution distribution coefficients. However, the ratio of the distribution coefficients of both minerals is nearly constant in relevant temperature regions [5.10]. Thus, a temperature determination from equation (5.16) has a much greater error than a temperature estimation made purely from the solubility of sylvite. In equation (5.14) the temperature-dependent distribution coefficients $C = C(T)$ are known from laboratory measurements; the concentration of bromine in ocean water (m_1) is also known. The bromine content at the core of the sylvite crystal yields the concentration (m_2) at the onset of crystallization. Fig. 5.4 shows the temperature dependence of the

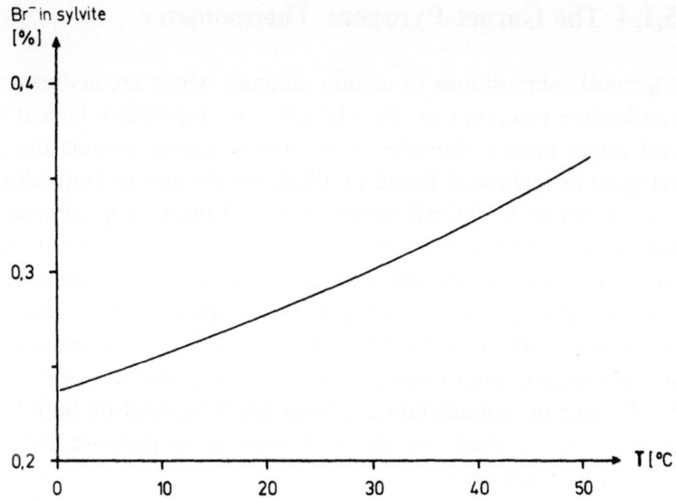

Fig. 5.4. Temperature dependence of bromine content in sylvite (KCl) after [5.71]

bromine content in sylvite. The ratio of the bromine content in sylvite and rocksalt is 10:1 and is not only dependent on temperature. Other factors include the composition and the depth of brine above the salt [5.7].

To obtain higher formation temperatures of certain ore deposits one utilizes the ion exchange properties of trace elements in crystal lattices and of various ores [5.6]. Only one example will be given here, namely the temperature-dependent substitution of zinc through manganese in wurtzite (ZnS) as well as lead with cadmium in galena (PbS).

Both minerals must belong to a parageneses such that they have been formed from a fluid phase of the same composition and at the same temperature. If these conditions are satisfied, equation (5.14) and (5.16) may be applied. If the ratio of the distribution constants is represented by R, then

$$\ln R = \ln(C_1/C_2) = \ln(m_1/m_2) = x_1/T + x_2. \tag{5.17}$$

The ratio C_1/C_2 is experimentally determinable so that the constants x_1, x_2 can be found. The temperature T can be obtained from the ratio of the mole fraction m_1/m_2, which can be determined analytically for the calculation of the formation temperature of wurtzite-galena paragenesis.

The determining equations [5.6] have the following form for trace elements

$$\text{Cadmium} \qquad T[^\circ K] = \frac{2580}{\ln R_{Cd} + 1.83} \tag{5.18}$$

$$\text{Manganese} \qquad T[^\circ K] = \frac{1890}{\ln R_{Mn} + 0.74}. \tag{5.19}$$

These equations are based not only on the assumption of simultaneous mineral formation of wurtzite and galena but also on the proportionality of mole fractions given in equation (5.14), valid only within a certain trace element concentration interval [5.71].

74

5.1.4 The Garnet-Pyroxene Thermometer

Chemical compositions of certain minerals which are assumed to originate from an equilibrium reaction can often be used as temperature indicators of the lower crust and upper mantle. Samples from greater depths include metamorphic rocks (e.g. eclogite) or inclusions found in alkali basalts and in kimberlites. The miscibility of various phases at defined temperature and pressure conditions is known from controlled laboratory experiments. By comparing these laboratory results with analytically obtained composition of natural rocks and individual mineral phases the formation conditions can be determined. Frequently used geothermometers are the diopside-enstatite-miscibility [5.49], the solubility of aluminum in enstatite [5.45] and the iron-magnesium-exchange reaction in the garnet-clinopyroxene-system [5.5, 5.35, 5.58]. Since the equilibrium reactions are influenced by both temperature and pressure, reactions which are the least sensitive to pressure are best used as geothermometers. These are:

$$Fe_3^{2+} Al_2[Si\ O_4]_3 + 3\ Mg\ Ca\ Si_2\ O_6 \rightleftharpoons$$
(Garnet)   (Clinopyroxene)

$$Mg_3\ Al_2[Si\ O_4]_3 + 3\ Fe^{2+}\ Ca\ Si_2\ O_6$$
(Garnet)   (Clinopyroxene).

In this reaction the $Fe-Mg$ ratio in garnet and pyroxene changes. The quotient of both is defined as the distribution coefficient (K_D):

$$K_D = \frac{(Fe^{2+}/Mg^{2+})_{garnet}}{(Fe^{2+}/Mg^{2+})_{pyroxene}}. \tag{5.20}$$

The temperature (T) dependence and pressure (p) dependence of the distribution coefficient is described approximately by [5.58]

$$K_D = f(T, p) = \frac{x_1}{T} + \frac{x_2}{R\,T}(p - p_0) + x_3 \tag{5.21}$$

with constants x_1, x_2, x_3 and universal gas constant (R). Equation (5.21) shows that the distribution coefficient decreases with increasing temperature and has the opposite behaviour for pressure.

Another approximation yields the partial derivation with pressure at constant temperature

$$\partial(\ln K_D)/\partial p = \Delta V/(R\,T) \tag{5.22}$$

with ΔV the volume change per mole. In the given garnet-clinopyroxene reaction the volume variation is small, so that the pressure influence is relatively small.

A further influence on the geothermometer, but not determinable analytically, is the chemical composition of the total rock in which garnet and clinopyroxene are present. The experimental investigations of most pure phases can be transfered to natural systems only with limitations. At present, investigations are inadequate for the correction of all natural samples to a normalized chemical composition of rock. The substitution for example of titanium for aluminum in clinopyroxene hardly influences the distribution coefficient, whereas the influence of larger contents of manganese or calcium cannot be ignored [5.35]. Experimental investigation of complex synthetic systems lead to a determining equation of the reaction temperature [5.58] which is valid for a broad interval of magnesium-iron ratio and directly applicable to rock consisting mainly of garnet and clinopyroxene.

If the distribution coefficient and pressure are known, the temperature and reaction equilibrium can be obtained from the determining equation

$$T[°C] = \frac{3686 + 0.2835\,p\,[MPa]}{\ln K_D + 2.33} - 273. \tag{5.23}$$

This equation is valid for magnesium-iron components in the range

$$0.062 < \frac{Mg^{2+}}{Mg^{2+} + Fe^{2+}} < 0.85$$

and can be applied directly to eclogite and garnet-pyroxenite.

On the other hand its application to ultrabasic inclusions in basalts is somewhat doubtful because the effects of olivine, orthopyroxene, spinel and amphibole on the distribution coefficient are not sufficiently well known [5.36].

The accuracy of the garnet-pyroxene thermometer is given at $\pm\,20\%$ for the pressure determination and $\pm\,5\%$ for the temperature determination [5.9]. The Fe$-$Mg and Ca$-$Mg exchange reactions between pyroxene and garnet exhibit a rather small dependence on pressure, so that pressure estimates based upon this small effect contain a large uncertainty of $\pm\,20\%$. In comparison the temperature determination provides an accuracy of about $\pm\,5\%$ [5.9, 5.52, 5.54]. The defining equation (5.23) is plotted in Fig. 5.5 using pressure as the independent variable.

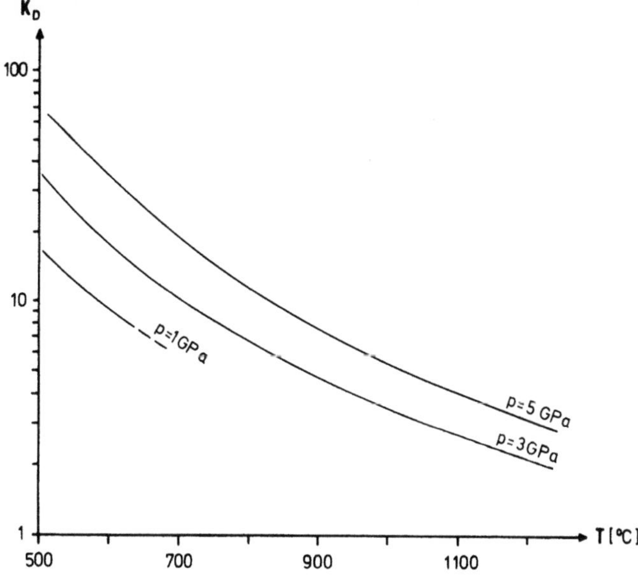

Fig. 5.5. Temperature and pressure dependence of the distribution coefficient K_D for the garnet-pyroxene thermometer after [5.58]

Another experimentally determined temperature and pressure dependence for a distribution coefficient has been applied to garnet-pyroxene inclusions in alkali basalts and on alpine type peridotites [5.35]. The low K_D-value obtained leads to high equilibrium temperatures (T $\leq$ 1100 °C) for the exchange reaction in these rocks.

Some caution should be shown in drawing conclusions about a general temperature distribution of the earth's interior from the inclusions in alkaline basalts and kimberlites. The xenoliths originated in thermally anomalous regions of magma formation and may not yield temperatures comparable to a regional temperature distribution in the mantle [5.36].

5.1.5 The Dolomite-Calcite Thermometer

In high temperature metamorphic terrains, a geothermometer has been developed based on the subsolidus phase relations in the binary system $CaCO_3 - CaMg(CO_3)_2$ [5.26, 5.31, 5.55, 5.56]. The magnesium content of calcite coexists with dolomite dependent on temperature as shown in Fig. 5.6. Since the dependence on pressure is weak, this system is favourable for temperature determinations. The slightly increasing solubility of $MgCO_3$ in calcite with pressure rises the solvus less than 0.05 °C/MPa [5.26]. This enables the application of the thermometer, for a rough estimation of the depth of metamorphism. Exsolution phenomena of dolomite in magnesium calcite are commonly observed in different mineral associations of high-grade metamorphic zones. The geothermometer can be applied in the temperature range between about 400 °C and 800 °C according to the reaction used. The dolomite content is analysed either by point counting or using the electron microprobe. A diffraction analysis provides the amount of unmixed proportion of $MgCO_3$ in calcite, so that the total

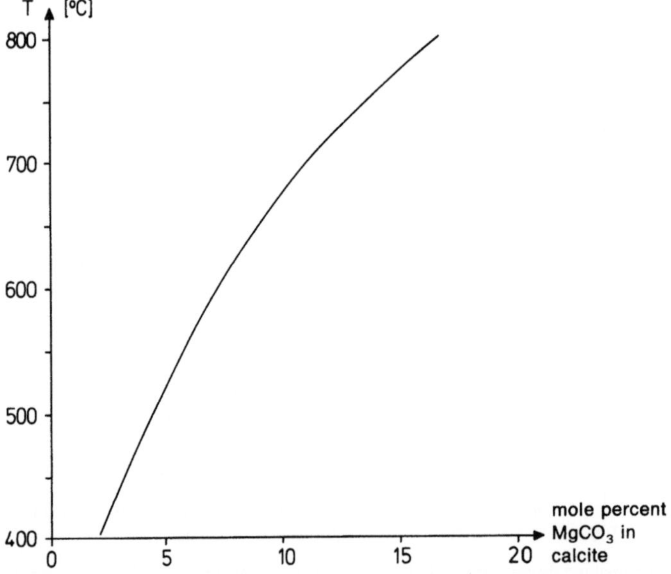

Fig. 5.6. Unique polybaric calcite-dolomite solvus after [5.30, 5.69]

mole percent $MgCO_3$ in calcite determines the temperature by applying the dolomite-calcite solvus in Fig. 5.6 [5.26].

5.1.6 The Degree of Coalification of Organic Inclusions in Sedimentary Rocks

Plant particles are commonly deposited together with the deposition of sediments. These particles are altered when the sediments subside. Initially most of the water is squeezed out because of the overburden pressure and accompanying mechanical effects. Then, after reaching even greater depth, i.e. about 500–1000 m, chemical changes take place which turn the plant component to brown coal. With progressively more subsidence lignite will be transformed into peat, anthracite and finally graphite. The organic particles are generally dispersed in the sedimentary rock. Only in exceptional cases do they accumulate such that coal seams form. The interpretation of coalification give an answer to various problems in geothermometry, in estimation of burial depths, and in reconstructing paleogeographical and tectonic history [5.68].

The different designations of coal are a qualitative measure of the degree of coalification of the particles. For a quantitative understanding of coalification, chemical criteria can be used [e.g. 5.46]. In this case the optical determination of the reflectivity of the organic substance has proved to be a very suitable measure of the coalification. Particles of size 10^{-3} mm are adequate for the determination and furthermore the method is non-destructive for the particles [5.66].

The influence of pressure on the organic substance is probably limited to mechanical packing of the material itself. Pressure sensitivity of the degree of coalification has not been absolutely proven. On the other hand, temperature has a significant influence on the progress of the chemical reactions governing coalification. The temperature dependence of the rate constant (k) in the modified Arrhenius equation amounts to

$$k = A \exp\left(\frac{1}{n} \sum_{i=1}^{n} \frac{E_i}{RT}\right) \tag{5.24}$$

with E_i $(i = 1, \ldots, n)$ the activation energies of the n possible reactions and R the universal gas constant.

Equation (5.24) shows that for low temperatures a much higher rate constant dominates in the course of a certain reaction than for higher temperatures. For a total unconverted mass of the same chemical composition (m_0) one can assume that after a time t only the mass (m) remains unconverted:

$$m = m_0 \exp(-kt). \tag{5.25}$$

Temperature and time are therefore the physical quantities which significantly influence coalification. For this reason the coalification process lends itself as an analytical method for the solution of geothermal problems [e.g. 5.14, 5.40, 5.44, 5.47, 5.69]. Various temperature-time functions can be deduced and plotted in relationship to the degree of coalification. The coalification temperature can generally not be obtained by direct measurement because in many cases it has changed since the termination of coalification. Examples are found in the Ruhr basin or the Bramsche massif. Only in

78

exceptional cases does coalification take place at a uniform temperature. Not only does a cooling intrusive body change the temperatures in the adjacent rock layers, but also the crustal uplift and subsidence contribute to continual changes in the reaction temperatures of the organic matter. The cooling off of a particle caused by uplift delays or may even stop the reaction, while during subsidence the constant rise in temperature causes increases in the degree of coalification. Because of the time dependence of the coalification reaction (equation 5.25) it is not possible to adapt quickly to a higher reaction temperature, so that the time duration (t) which an organic particle has spent under specific temperature conditions is important.

It is known from experience that the degree of coalification increases with both rising temperature and time of its exposure, but an analytical concept is lacking, because of the most complex chemical composition of the organic matter. As a first approximation, it is assumed that the mean optical reflectivity (R_m) of vitrinite, resp. humite under oil, which is a common measure of the degree of coalification, is proportional to the integrated value of temperature and time of its exposure for progressive subsidence of sedimentary layers:

$$R_m \sim \int T(t)\,dt. \tag{5.26}$$

If the temperature gradient within a borehole or at least within a given depth interval is taken as constant, the integral value can be calculated from a subsidence curve:

$$\int T(t)\,dt \approx dT/dz \int z(t)\,dt. \tag{5.27}$$

Correlation tests have shown that the square of the reflectivity (R_m) of vitrinite or humite under oil is proportional to the temperature gradient and the subsidence history [5.13, 5.14].

$$R_m^2 = f(dT/dz) \int_0^{t_1} z(t)\,dt. \tag{5.28}$$

Under a microscope vitrinite and humite are distinguishable mazeral groups of the humous components. The mean reflectivity lies within the interval $0.2 < R_m < 5\%$. For graphite, which has the maximum degree of coalification this can be even higher.

The integral of subsidence history of a particular organic particle down to a depth of $z = z(t_1)$ from which the sample for the reflectivity measurement is taken, is calculated from borehole data. The subsidence curve is constructed from the borehole profile.

Often the subsidence history is not fully known for the entire duration but only for a time interval $\hat{t} \leq t \leq t_1$, as for example in the Ruhr basin. In such a case, the time integration begins with time $\hat{t}$, from which point the sequence of subsidence is known.

The integral over the time $0 \leq t \leq \hat{t}$ is however a constant (c) in which case equation (5.28) takes on the following form:

$$R_m^2 = f(dT/dz) \left[c + \int_{\hat{t}}^{t_1} z(t)\,dt \right]. \tag{5.29}$$

The constant c calculated for a specific borehole can of course include further influences on the rate of coalification

- temperature and time influence during uplift and further subsidence
- long periods of inactivity in which at first the degree of coalification
  increases at suitable depths, but soon reaches maximum attainable degree
  of coalification at a certain temperature.

Beside the subsidence history as a variable in equation (5.28) the calibration function $f = f(dT/dz)$ is a second variable to be determined. For the sake of simplification the calibration function is considered in the sectional plane as $R_m = 1.0\%$ and the value of the integral for this reflectivity is called I:

$$I = \int_0^{t_1} z(t)\, dt\big|_{R_m = 1}.$$

(5.30)

For suitable boreholes with known temperature gradients and determined I-values the determining equation for the temperature gradient [5.14] is

$$dT/dz[^\circ C/km] = 98.7 - 14.6 \ln I\,[km\ Ma].$$

(5.31)

Use of these equations on the well known coalification pattern in the middle Upper Rhinegraben yields a geothermal history for this region (Fig. 5.7) which can be compared to the volcanic activity of the Rhinegraben and the bordering areas [5.13, 5.14, 5.15]. In the early thermal history of the Rhinegraben, prior to the recent geothermal anomaly, a maximum in temperature gradients ($dT/dz = 70\ ^\circ C/km$) existed in the time span of the rift formation in Eocene to the Lower Oligocene. Afterwards the temperature gradient sank to a lower level ($dT/dz \approx 40-50\ ^\circ C/km$). The

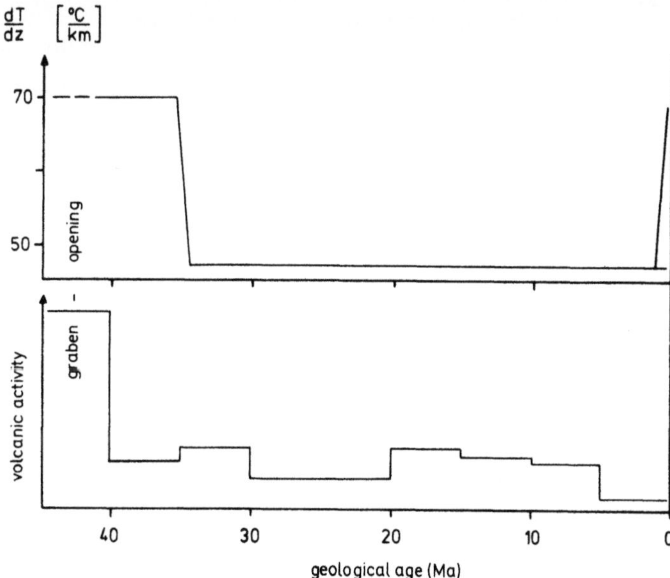

Fig. 5.7. Geothermal history of the middle Upper Rhinegraben (top) and volcanic activity in the Upper Rhinegraben region (bottom) after [5.13]

presently higher gradient has as yet not appreciably influenced the coalification ratio in the Upper Rhinegraben [5.13, 5.14, 5.67].

The degree of coalification always reflects the maximum temperature after sufficient time of influence. A time change in the temperature gradient during subsidence becomes visible in a coalification profile only if the temperature gradient of the past has been greater than subsequent gradients. Therefore, it is impossible to give information as to the initial increase in temperature gradients for the Rhinegraben.

A further example of the thermal evolution of an area where there is a heat flow density anomaly is the Urach region [5.16]. The available coaly organic particles in the sedimentary rock of the Jurassic, Triassic, Permian and Carboniferous time are not only considerably older, but also include a far larger geological time period than the Tertiary of the Upper Rhinegraben. The calculated temperature gradient of $dT/dz = 43$ °C/km probably influenced the coalification condition during the largest subsidence of layers during Malm time. The high temperature gradient may still have dominated during Cretaceous time and in the early Tertiary before uplift began. The present temperature gradient of $dT/dz = 45$ °C/km [5.73] corresponds to the paleo-geothermal gradients. Yet it is difficult to conceive that the temperature gradient has remained constant over such a long period and above all survived the uplift phase. It is however possible that the origin of the heat source is very deep and therefore when it is traced to the surface is very constant in time. An accompanying symptom of such deep seated heat anomaly could be the basalt vents of the Urach region.

The correlation of the present high heat flow density of the Urach and the paleo-heat flow density of this region leads one to believe that the Swabian anomaly and the Upper Rhinegraben have the same deep-seated root. In this case the exact dating of the high paleo-geothermal gradient is possible. This high heat flow density perhaps already dominated at the beginning of the subsidence. In analogy with the Upper Rhinegraben, the Urach anomaly in the early Tertiary may have faded and more recently fortuitously increased to its old level.

Paleogeothermal gradients in sedimentary basins of different geological ages are reported. The Carboniferous Ruhr basin represents a high thermal regime with gradients of about 70 °C/km during Westphalian [5.17]. Permian basins of Australia show gradients somewhat lower, i.e. 40 to 50 °C/km [5.48]. The measured temperature gradient in Tertiary basins, however, vary over a wide range, so that no conclusion can be drawn from these examples about the thermal evolution of the crust.

5.2 Geophysical Methods for Determining Temperatures

The thermometer can only be used as a measuring device for temperatures in the earth in the accessible uppermost regions of the crust. Depending on the problem, a mercury thermometer, thermocouple, thermistor or a pyrometer may be used. For inaccessible regions indirect methods have to be employed. In order to obtain these temperatures one uses the temperature dependence of physical properties of rocks such as density, electric conductivity or velocity of sound. Precise measurements of these quantities and their lateral as well as vertical variations within the interior of the earth yield indirect information about the thermal state of the earth.

5.2.1 Direct Measurements at the Earth's Surface and in Borehole

The temperature of the earth's surface can be determined as a radiation temperature by applying Stephan-Boltzmann's law (Sect. 4.1.1) to the observed spectrum of infrared light. Such infrared measurements can be made from satellites and airplanes with a so-called IR-scanner [5.28, 5.60]. These measurements are only rarely used to determine average soil temperatures but are used, for example, in the dense jungle of South America to map rivers or to detect the mixing of water flows of different temperatures. In a few cases, where thermal water reaches the surface and discharges in appreciable amounts, a genuine geothermal system anomaly can be recognized. However in normal circumstances the influences of topography, vegetation and solar radiation are so great that the background heat flow density, which amounts to only about $10^{-3}\%$ of the solar constant (Sects. 4.1.1 and 4.1.5) can vary over a broad interval, within changing surface temperatures appreciably.

Because of the surface conditions which change rapidly in the course of a day, infrared measurements are unsuitable for determining or estimating the relative time-independent soil temperatures that exist at a depth of a few centimeters or meters depending on the application. For soil temperature measurements, probes must be used which register the time temperature changes in order to make the proper corrections. If the annual temperature variation can be ignored (i.e. if the soil temperatures are to be obtained within a time interval of few days) or only temperature differences are to be determined then it is sufficient to use probes which measure the daily variation down to a depth of about $x = 25$ cm. Mapping soil temperatures in the uppermost meter using corrected daily variations can expose temperature anomalies which are of importance for hydrological problems. For example, one can detect anomalies which are related to the vertical discharge of thermal waters and also those that originate from the flow of warm water at depth.

Soil temperatures can be measured using probes outfitted, for example, with resistance thermometers. Because the amplitude of temperature variations decreases exponentially with depth in the uppermost regions beneath the surface (Sect. 4.1.1.1), temperatures near the surface are determined over smaller depth intervals than at deeper regions. Such a probe is illustrated schematically in Fig. 5.8. Each resistance thermometer is part of an electric bridge circuit in which a small deviation of the bridge voltage from the null position is proportional to temperature. The soil stationary temperature can be deduced from the daily variation of temperature (Fig. 5.9) using equations (4.4) and (4.5).

At a depth of $z = 50$ cm the amplitude of the daily cycle has almost disappeared. This can be seen from equation (4.5) whereby $T_{max}(50)/T_0 < 0.01$. Thus temperatures at this depth are obtained simply if the probe can penetrate the soil easily. For greater depths of temperature measurement, a hole must be augered or drilled and the difficulty in drilling increases with depth. In order to obtain ground temperatures which are independent of the annual temperature cycle, boreholes must be about $10-15$ m deep [5.43] and temperatures must be determined at various depth from the surface.

With still deeper boreholes, the disturbance of the rock temperatures due to the drilling process itself becomes more evident. Water circulation from the drill rig and heat produced at the drill tip generally warm up the rock adjacent to the borehole.

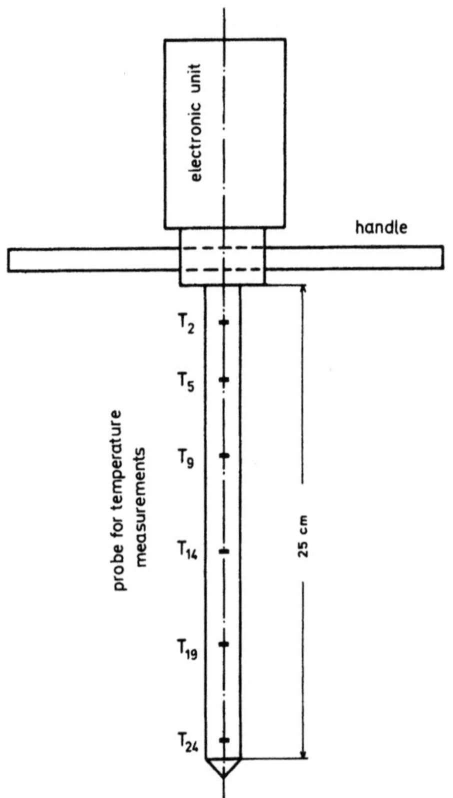

Fig. 5.8. Schematic representation of a probe which can register diurnal temperature variations within the soil layer

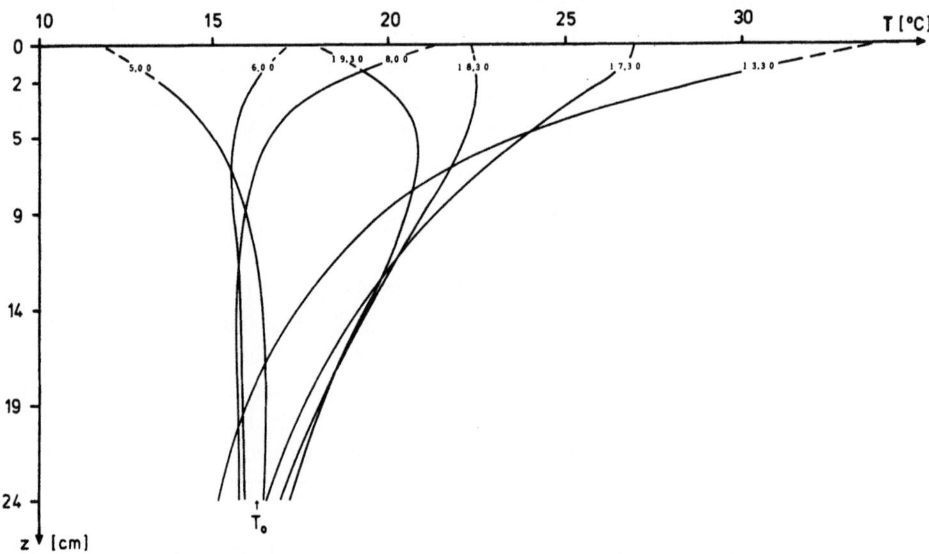

Fig. 5.9. Diurnal variations in temperature and their dependence on depth (14/15. 8. 79, Clausthal-Zellerfeld)

However, depending on the magnitude of the temperature gradient at a particular drill locality, the relationship of the circulation temperature and the rock temperature can reverse so that below a certain depth the rock is cooled off. At the same time the circulation water is heated and then releases a part of the heat to the upper borehole region. Thus a transient disturbed temperature profile results which is schematically shown in Fig. 5.10 together with an undisturbed temperature profile.

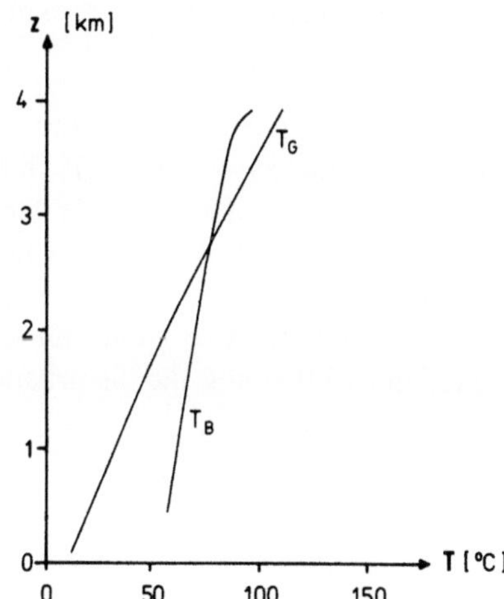

Fig. 5.10. Schematic representation of a disturbed borehole temperature measurement (T_B) and an undisturbed temperature-depth distribution (T_G) after [5.32]

The original temperature field is restored only after a very long time, and a correction of the measured values is not promising because the disturbance itself is not a well defined analytically described process. Often the circulation is interrupted for an undetermined time in order to change the drill bit or to set casing. Furthermore when ground water penetrates the borehole or when circulation fluid is lost, uncontrolled temperature changes occur in the rock. These are very difficult to correct. The time necessary to attain an approximate thermal equilibrium can be months. It is primarily dependent on the circulation time, the borehole diameter and the temperature difference between circulation fluid and adjacent rock.

An empirical method for determining rock temperature soon after drilling is based on a graphical analysis of the transient temperature curves [5.32]. For this purpose temperatures at the bottom of the borehole are measured at certain intervals following termination of circulation (e.g. using a mercury maximum thermometer or some other suitable device). Temperatures return slowly to their undisturbed equilibrium values, and so the asymptotic attainable temperatures can be determined graphically. This method has the disadvantage that temperature measurements are made during standby time for the drill rig, which means that the measurements have to be accomplished within a day or so after termination of drilling. However, this time is so far removed from the time required for equilibrium temperatures to be attained, that rock temperatures are extrapolated with great uncertainty.

A mathematical method for calculating rock temperatures in boreholes is given by [5.11]. The borehole is considered a linear heat source (Q), causing a temperature

84

disturbance ΔT at a distance r from the borehole axis

$$\Delta T = \frac{Q}{4\pi K} \int_{x_1}^{\infty} \frac{\exp(-x)}{x}\, dx \quad \text{with} \quad x_1 = \frac{r^2}{4\kappa t}\, 0 \le t \le t_1. \tag{5.32}$$

K and κ are the thermal conductivity and thermal diffusivity of the rock respectively.

After the termination of circulation at time $t = t_1$ a heat sink $(-Q)$ of equal strength is added, so that for time $t \ge t_1$, the following disturbance is produced:

$$\frac{\Delta T}{T_0} = \left(\int_{x_1}^{\infty} \frac{\exp(-x)}{x}\, dx - \int_{x_2}^{\infty} \frac{\exp(-x)}{x}\, dx \right) \Big/ \int_{x_3}^{\infty} \frac{\exp(-x)}{x}\, dx \tag{5.33}$$

where T_0 is temperature at the borehole wall $(r = a)$ at time $t = t_1$, $x_2 = r^2/(4\kappa(t - t_1))$ and $x_3 = a^2/(4\kappa t_1)$. This equation is not valid in the bottom region of the borehole because of the infinite line source assumption. It can be shown that the disturbance penetrates only a few meters into the wall rock depending on the size of individual parameters.

For long times $x_i \ll 1$ one can substitute the following approximation [5.11] for equation (5.33) giving the disturbance at the borehole wall $(r = a)$:

$$\frac{\Delta T}{T_0} = \ln\left(\frac{t}{t - t_1}\right) \Big/ (\ln(4\kappa t/a^2) - 0.577). \tag{5.34}$$

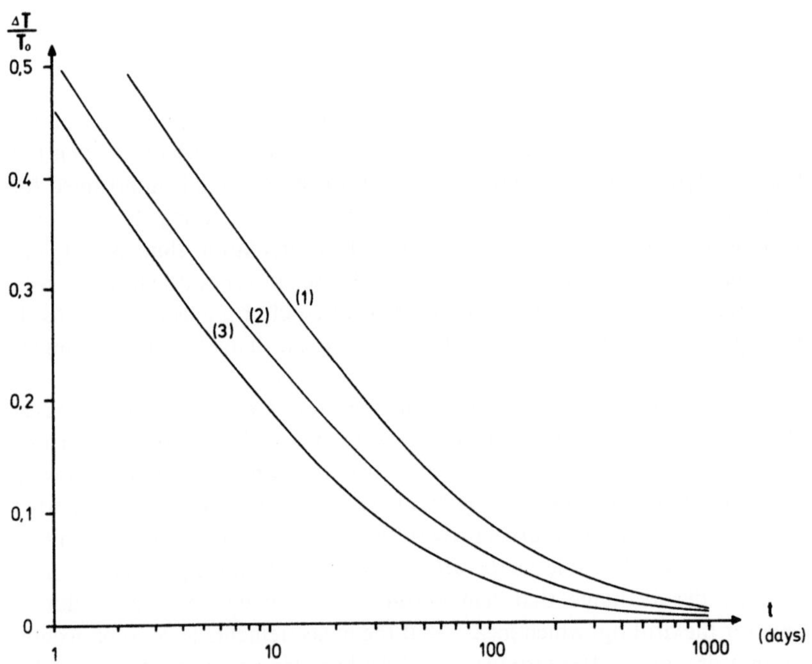

Fig. 5.11. Relative change of temperature at the borehole wall following completion of the drilling process $(t = 0)$ for these examples
1) Borehole radius a = 10 cm; duration of drilling t_1 = 90 days
2) a = 6 cm, t_1 = 60 days
3) a = 6 cm, t_1 = 30 days

The ratio $\Delta T/T_0$ is plotted in Fig. 5.11 against time $t > t_1$ for boreholes with

1) $a = 10$ cm, $t_1 = 3$ months
2) $a = 6$ cm, $t_1 = 2$ months
3) $a = 6$ cm, $t_1 = 1$ month

and a thermal diffusivity $\kappa = 10^{-6}\, \text{m}^2/\text{s}$.

Equation (5.34) can also be used in the following form without measuring the temperature at the borehole wall immediately after termination of water circulation [5.18, 5.42]

$$\Delta T = Q/(4\pi K)\, \ln(t/(t - t_1)). \tag{5.35}$$

Frequently temperatures in boreholes are measured only at the bottom of the holes after technical interruptions. As a rule, drilling is continued in one or two days, limiting the time for determining undisturbed rock temperatures. One method of determining temperature estimation based on an exponentially decreasing disturbance [5.2] is:

$$T_\infty^i - T(t) = (T_\infty^i - T_0^i)\, \exp(-c^i(t - t_0^i)). \tag{5.36}$$

After a point in time $t = t_0^i$, which marks the beginning of an i-th time interval, several temperatures $T = T(t)$ are measured during a specific time and thereby T_∞^i as well as c^i are estimated. The c^i converge and go to zero in a stationary condition. The boundary value of temperature (T_∞) can be determined graphically if T_∞^i is plotted against c^i thereby determining the undisturbed rock temperature for $c = 0$.

5.2.2 Indirect Methods for Determining Temperatures

The accuracy in geophysical measuring techniques as well as interpretation techniques for determining petrophysical quantities of the earth's interior have steadily improved over the last few decades. These techniques, combined with a knowledge of the temperature sensitivity of certain physical parameters, have helped establish a temperature field for the earth. For example gravimetry methods provide models for density distribution including possible inhomogeneities which may be temperature-dependent [e.g. 5.12, 5.38]. The electric conductivity of sedimentary rock as well as crystalline rock is temperature dependent. Thus, geoelectric and magnetotelluric methods which yield a distribution of specific resistance also provide information about temperature anomalies [e.g. 5.34, 5.37]. Another physical property used as a temperature indicator of the earth's interior [5.30] is the fading of remanent magnetization at the Curie-temperature. Finally with the onset of melting or with solid state phase changes seismic velocity decreases sufficiently such that a recognized velocity gradient can possibly be interpreted as temperature effect.

None of these indirect methods taken alone will yield temperature distribution of the earth's interior with certainty. Most physical parameters are very material-dependent so that the effects of temperature play a subordinate role. Only by studying a number of results can the temperature effects be obtained.

5.2.2.1 Temperature Determination from Gravimetric Measurements

In gravimetry local changes in the gravity field are interpreted as arising from density variations within a basic model. The origin of such density differences could be material or structure related (e.g. a change in the rock porosity, crack density, or by fault zones) or could be related to a temperature field. When using gravimetry as an indirect measuring method to explore for temperature differences in the subsurface it is not possible to determine "a priori" a definite temperature distribution from the shape of the gravity field. Gravimetry can be used only if detailed knowledge of the geological structure is available and if one can roughly localize thermal anomalies. With these assumptions a quantitative estimation of temperature distribution can be made. Thermal anomalies with surface expressions such as steam or thermal springs are frequently associated with negative Bouguer anomalies [5.27, 5.51]. The cause of such minima can vary. Even if one assumes nearly homogeneous material in the subsurface, porosity differences within the fault zone can enable thermal water to rise from great depths. Horst structures which have layers of higher porosity also serve as storage locations of hot water. Thermal water rising from deeper layers heats up the rock through which it passes, decreasing its density and thereby increasing the amount of the negative Bouguer anomaly. The measured negative gravity anomaly (Δg) is made up of two parts:

$$\Delta g = \Delta g_{th} + \Delta g_{por} \tag{5.37}$$

the temperature-related anomaly (Δg_{th}) and that which is generated because of higher porosity (Δg_{por}). The part Δg_{th} contains the temperature-dependent term, because the gravity anomaly is directly proportional to a density change ($\Delta \varrho$)

$$\Delta g_{th} = \gamma \, \Delta \varrho \, f(\check{x}) \tag{5.38}$$

where $\gamma = 6.67 \cdot 10^{-5} \, \text{mgal cm}^2 \, \text{g}^{-1}$ is universal gravitational constant and $f = f(\check{x})$ is the geometric form factor which depends on the geometry of the anomalous body and the point at which gravity is being considered.

In a unit of rock of density ϱ_0 a temperature increase from $T = T_0$ to $T = T_1$ leads to the density decrease

$$\Delta \varrho = \varrho_0 \, \alpha \, (T_1 - T_0) = \varrho_0 \, \alpha \, \Delta T \tag{5.39}$$

with a volume expansion coefficient α.

From equation (5.38) and (5.39) one obtains the relationship:

$$\Delta T = \Delta g_{th} (\gamma \, \varrho_0 \, \alpha \, f(\check{x}))^{-1}. \tag{5.40}$$

Thus the calculation of temperature differences from gravity studies still requires assumptions concerning the partitioning of a thermal component of the gravity anomaly as well as about the density distribution. As a consequence knowledge of geological structure is still essential. Where possible information about the geological structure and temperature distribution should be obtained from boreholes.

The division of a gravity anomaly into its thermal and structural origins after equation (5.37) is difficult. It has been shown from known temperature anomalies that the thermal portion contributes about half of the determined anomaly Δg.

In order to explain thermally caused gravity anomalies of a regional nature which cover large crustal regions and the upper mantle, local near-surface density inhomogeneities are unimportant. In this case the total observed gravity anomaly can be interpreted as having a thermal origin. In contrast to local thermal gravity anomalies which reach only $\Delta g \approx - 1$ mgal with local temperature anomalies, regional thermal gravity anomalies can be considerably greater because of much larger deep temperature differences [5.38].

It should also be remembered that the Bouguer anomaly is sensitive to density changes along specific horizontal layers. With large changes in topography there exists horizontal temperature difference down to depths of several kilometers. The temperature differences at sea level in the case of the Alps amount to $\Delta T \sim 60\ ^\circ C$ between the Central Alps and the Alpine foreland, thereby giving rise to a regional thermal gravity anomaly of $\Delta g \approx - 6$ mgal [5.12].

5.2.2.2 Temperature Determination from Geoelectric Measurements

The subdiscipline of geoelectrics is concerned principally with determining differences in specific resistance within the earth. Thus geoelectric methods can be used to locate boundaries of various rock types and tectonic disturbance zones and isolate saline and freshwater layers.

The specific resistance for dry rock is high, but can very over 13 orders of magnitude. It lies within the range of $10^5 < \varrho < 10^{18}$ Ωm. In contrast electrolyte has a relatively small value of a few hundred ohm meters. The specific resistance of a water-saturated porous rock is therefore determined essentially by the specific resistance of electrolytes and the porosity of rock.

An experimentally determined relationship between the specific resistance (ϱ) and the porosity (Φ) of a rock is

$$\varrho = \varrho_E\ \Phi^{-m}\ S^{-2} \tag{5.41}$$

where ϱ_E is the specific resistance of the electrolyte, S ($0 \leq S \leq 1$) is the degree of saturation and m is a constant. In the case of loose grains $m = 1$ and with strongly cemented sedimentary rock m reaches a value of 2.2. This equation is no longer valid for small porosities because then surface conductivity (σ_q) which is neglected for larger porosities [5.59] dominates.

The specific conductivity σ of rock is given by

$$\sigma = \sigma_E/F + \sigma_q \tag{5.42}$$

with the formation factor F and the electrolyte conductivity σ_E. The formation factor includes effects of porosity, pore structure, size distribution of pores and clay fraction.

The temperature dependence of electrical conductivity is based solely on the temperature-dependent ionic conductivity up to the boiling point of the electrolyte. Thus the increase in conductivity of water-saturated porous rock with rising temperature can be expressed in terms of the temperature coefficient (α) of the ionic conductivity:

$$\sigma_E = \sigma_0(1 + \alpha\ \Delta T). \tag{5.43}$$

88

The term ΔT is the temperature difference between $T = T_0$ and $T = T_1$ causing the increase in conductivity.

Equation (5.43) can be combined with equation (5.42) yielding

$$\sigma = \sigma_0(1 + \alpha \, \Delta T)/F + \sigma_q. \tag{5.44}$$

Often instead of the formation factor F only the porosity Φ is given so that in the case of saturated rock (S = 1), the following temperature dependence results from equation (5.41):

$$\sigma = \sigma_0(1 + \alpha \, \Delta T) \, \Phi^m. \tag{5.45}$$

The electric conductivity σ_0 is determined by the specific ionic composition of the pore fluid. As a rule NaCl concentration is much higher than that of other salts present. Thus σ_0 can be determined from experimental results based on the salinity of NaCl solutions [5.57].

In order to estimate temperature from conductivity measurements in a geothermal region, the salinity as well as porosity or formation factor must be known. The additional necessary parameters can only be obtained from boreholes. Thus in a region under investigation it is necessary to drill at least one test borehole before a temperature distribution can be calculated by using the relationships given in equations (5.44) and (5.45).

The temperature coefficient is not a constant, but is itself a function of temperature $\alpha = \alpha(T)$. For the first approximation, based on experimental results [5.57] the temperature coefficient can be linearized in 2 intervals:

$$\alpha = \begin{cases} 0.026 \, ^\circ C^{-1} & \text{for} \quad 0 \leqq T \leqq 200 \, ^\circ C \\ 0.001 \, ^\circ C^{-1} & \text{for} \quad 200 < T \leqq 350 \, ^\circ C. \end{cases}$$

Since the temperature sensitivity of conductivity above $T = 200 \, ^\circ C$ is slight, inaccuracies in temperature determination increase in this range.

The pressure dependence of electrolyte conductivity is small and can be ignored up to pressures of $p = 10$ MPa ($= 100$ bar). Under hydrostatic conditions this pressure is reached at a depth of $z = 1000$ m and under lithostatic pressure at a depth of $z = 400$ m. In many geothermal systems the pressure in the pore fluid is close to hydrostatic pressure, therefore temperature determinations from distribution of specific resistance or conductivity can be determined to $z = 1000$ m depth without considering pressure effects.

5.2.2.3 Results from Magnetotellurics as Temperature Indicators

Natural electromagnetic fields, especially in the frequency range below 1 Hz used in magnetotellurics, penetrate deep into the earth's crust and into the upper mantle. Higher frequencies are limited to shallower penetration because of the skin effect. Using measured field components interpretation techniques of magnetotellurics permit the construction of a model of specific resistance within the earth. Dry rocks are generally non-conductors at low temperature. However, lattice defects within the crystal structure as well as impurity ions cause a small electron and ionic conductivity which increases with temperature. Experiments indicate that semiconduction condition is reached between T = 800 and 1200 °C. The temperature dependence of specific conductivity (σ) is described by

$$\sigma(T) = \sigma_0 \exp(-E/kT) \tag{5.46}$$

with E an average activation energy, k Boltzmann's constant, T absolute temperature and σ_0 the extrapolated conductivity for T → ∞. The quantities E and σ_0 are only constant within certain temperature intervals, because the conduction mechanism varies. In lower temperature regions conduction caused by crystal defects is dominant, whereas at higher temperatures intrinsic conduction (Fig. 5.12) dominates. High pressures increase conductivity slightly [e.g. 5.72].

From the vertical distribution of the specific resistance one cannot make firm conclusions about a temperature distribution. In the upper crust the water content in pore spaces influences the conductivity so strongly that the matrix conductivity can be ignored (see Sect. 5.2.2.2). For the deeper interior the material-dependent parameters E and σ_0 are not known well enough to be able to calculate a temperature distribution from magnetotelluric deep sounding.

In certain regions and with adequate knowledge of the parameters E and σ_0 it may be possible to bring temperature distribution derived from model calculations and results from magnetotelluric sounding into agreement [5.1]. Even then it is generally only possible to make this correlation for the case of deep-lying layers with low specific resistance. If these layers are localized within the upper mantle, then partial melting can be treated as ionic conductivity. In the case of a homogeneous solid phase it can be treated with semiconductor properties as in equation (5.46). Highly conducting layers in the crust can be explained by ionic conduction in a partial melt. Melting of granitic material begins at T = 600 °C, and basaltic rock at about T = 900 °C.

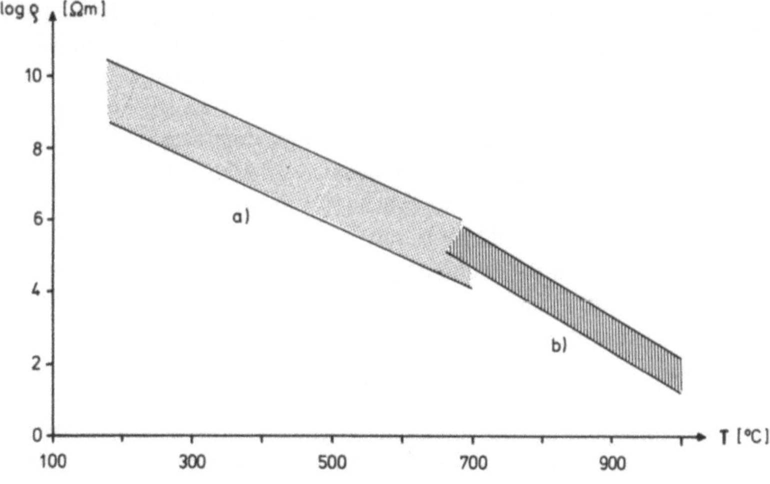

Fig. 5.12. Temperature dependence of the specific resistance of crystalline rock (schematic). Conduction is dominated in region (a) because of defects in the crystal lattice, in region (b) by intrinsic conduction of electrons after [5.72]

The interpretation of an electrical conductivity anomaly as a temperature effect [5.61] leads to the view that a change in the depth of highly conductive layer can be associated with vertical displacements of isotherms. The correlation between regional heat flow density and depth to highly conducting layers confirms this assumption [e.g. 5.1].

5.2.2.4 Shape of the Curie-Point as an Isotherm

Magnetite (Fe_3O_4) commonly occurs in the earth's crust in many crystalline rocks as an accessory mineral. Because of its magnetic properties it is classified as ferrimagnetic, to which almost all metal oxides, e.g. titanomagnetite and metal sulfide belong. Ferrimagnetism results from interactions of the magnetic moments of ions in crystal lattices [e.g. 5.3].

The regulating interactions are superimposed by thermal movement of ions when the temperature increases. The temperature at which the ferrimagnetic property finally disappears is called Curie-temperature. The Curie-temperature of magnetite is about $T_c = 580$ °C and depnds on the amount of titanium ions substituting for iron ions in the crystal lattice. With crystalline rocks a Curie-temperature of about $T_c = 550$ °C is measured [5.8]. However, ferrimagnetism disappears gradually rather than suddenly. At $T = 400$ °C a typical crystalline rock still maintains 70% of its ferrimagnetic property [5.8].

The lower surface of the magnetic layer of the crust is generally not a material boundary but a temperature boundary. Therefore changes in its thickness can be interpreted as variations in depth of the surface with temperature $T = T_c$. Regions with a Curie surface situated higher than its surroundings are detected on the surface by negative magnetic anomalies, which can be indicators for comparatively higher heat flow density within the area.

The interpretation of the negative anomaly, possibly produced by thermally induced changes in thickness of the magnetic layer, can be accomplished by spectral analysis of the measured anomalous magnetic field within a specific area [e.g. 5.33, 5.62]. The amplitude-frequence-spectrum obtained through Fourier analysis yields a first estimation of an average depth to the lower boundary of the magnetic layer. The amplitudes for these depths are independent of frequency.

A study of the Upper Rhinegraben [5.8] showed that the lower boundary of the magnetic layer is at $z = 8$ km. If this boundary is interpreted as the depth of the Curie-isotherm, then the average temperature gradient can fall between the interval $50 < dT/dz < 70$ °C/km, if $400 < T_c < 550$ °C. However if the bottom of the magnetic layer is interpreted as the boundary between magnetic and nonmagnetic rock cooler than the Curie-temperature, then temperature gradient data cannot be obtained.

Upwarping of the Curie-shape can be detected in the Urach region in the Swabian Alb, Germany. The shape of the Curie-temperature which occurs at about $z = 25$ km in regions having global average temperature gradients, is upwarped to 15 km at Urach [5.8]. The corresponding temperature gradient is increased to a value of $40 < dT/dz < 55$ °C/km.

5.2.2.5 Temperature Estimates from Seismic Results

Two characteristic quantities in rocks governing the propagation of seismic waves are velocity and absorption. Both of these properties are not only dependent on the specific material but also on temperature. Above all the compressional wave velocity decreases with increasing temperature [e.g. 5.20]. The shear wave velocity within rock is less sensitive for temperatures below the melting point. However, at the onset of partial melting it decreases rapidly and disappears totally when the cohesion of the crystal matrix is lost. The absorption coefficient

$$\alpha = \frac{1}{\Delta x} \ln(A_1/A_2) \tag{5.47}$$

which is calculated from the damping of the amplitude (A_1/A_2) of waves and the travel path Δx. The temperature region below the melting point is less temperature-dependent. In contrast, during fluid phases seismic waves become strongly absorbed [e.g. 5.63]. Thus from absorption and shear velocity one can determine the possible onset of melting within the earth or one can localize a possible magma chamber. Temperature determination is only possible with a large temperature interval.

For quarzite the compressional wave velocity (v_p) is above all very temperature-sensitive. The v_p-velocity in the vicinity of the phase transformation point from alpha-quartz to beta-quartz-modification at $T = 573\ °C$ is decreased by 15% in comparison to the v_p-velocity at $T = 100\ °C$. Thus, for granites containing 30% quartz one can expect velocity decrease of 5%, without considering the other rock-forming minerals. The temperature sensitivity of v_p-velocity increases considerably for granite at higher temperatures and reaches a maximum before the phase changes. Because the phase change is also pressure dependent (see Sect. 2.1.1) the temperature of maximum sensitivity increases correspondingly. Table 5.1 gives the temperature sensitivities β of v_p-velocities for three common rock groups based on measurements [5.20] at constant pressure

$$\beta = (v_p(0) - v_p(T))/(v_p(0)\,\Delta T)\ [°C^{-1}]. \tag{5.48}$$

Table 5.1 also exhibits temperature sensitivity β that on the one hand increases with rising temperature and on the other decreases with SiO_2 content of the rock.

Table 5.1. Temperature sensitivity (β) of the v_p-velocity of rocks after [5.20]

T [°C]	$\beta\ [10^{-4}\ °C^{-1}]$		
	Granite	Gabbro	Peridotite
200	0,6	0,5	0,4
300	0,8	0,6	0,55
400	1,1	0,75	0,6
500	1,55	0,9	0,6
600	–	1,25	0,75
700	–	(1,65)	(0,8)

Supplementary Problems

5.1 Ascending thermal water is heavily mixed with salt water in marine deposited layers, so that the original sodium content of 300 ppm increases up to 2000 ppm. The calcium content (1%) and potassium content (2 ppm) are unaltered. Determine the equilibrium temperature where the thermal water originates and the apparent temperature after contamination.

5.2 The optical reflectivity of an organic particle shows a value of $R_m = 0.4\%$. The sample comes from a depth of $z = 1500$ m in a continuously subsiding basin, and the age of the sample is $t = 20$ Ma. Estimate the temperature gradient in the basin using Eq. 5.26.

5.3 Two samples are taken from a coal mine in the Ruhr district. The reflectivities of the samples are $R_m = 1.35$ and 1.8%; the total depths, i.e. maximum burial 2.14 km and 2.69 km; the absolute ages are 309.2 Ma and 311.6 Ma. For integration and determination of the value c, the relative ages $t_1 = 0$ and $t_2 = 2.4$ Ma should be used. Estimate the paleotemperature gradient using Eq. 5.27.

5.4 During an interruption in drilling operations the following bottom-hole temperatures were measured $T_i = 92, 93.5, 95, 96, 97, 98\ °C$ at the time $t_i = 200, 268, 404, 500, 656, 968$ minutes after the interruption. Determine the equilibrium rock temperature using Eq. 5.34.

6 Geothermal Heat as an Energy Source

With recent energy shortages the possible use of geothermal energy as an alternative source has received much new attention. Geothermal energy, in the broadest sense of the term, is relatively evenly distributed in the earth. But it is economically feasible to harness this energy only when conditions are favorable. This is only true for volcanically young areas where hot steam can be found close to the surface and can be economically converted into electric energy. However, rising costs of energy make it possible to use smaller as well as more deeply situated heat reservoirs and with this the incentive to find smaller and deeper heat sources has increased.

Geochemical and geological as well as geophysical methods of prospecting for thermal reservoirs are employed. It is common to start with a reconnaissance exploration program. This may involve obtaining an initial overview from the air, e.g. infrared photography of the surface. Then a choice within anomalous thermal areas is made using a number of surface prospecting methods. Finally, the structure of the subsurface in the chosen region is determined by geophysical methods. Geophysical and geochemical results, together with test boreholes, are used to estimate the potential size of the heat source.

How a thermal reservoir will be utilized is determined mainly by its water temperature. When steam is produced, thermal energy can be converted into electricity. Hot water reservoirs alone are generally used for space heating of homes, swimming pools and greenhouses. If the heat is stored in hot dry crystalline rock, then thermal energy is obtained by inducing fractures at depth and extracting the heat by means of forcing water through the open cracks. Water trapped in deep reservoirs flashes to steam upon release of pressure and can be used at the surface to produce electricity or heat.

The use of thermal springs for medical purposes must not be underestimated as it contributes to the health, strength and economy of people.

6.1 Prospecting Methods for Thermal Reservoirs

There is a close connection between methods of determining temperatures within the earth's interior (Chapt. 5) and prospecting for thermal reservoirs. All methods used to determine present or recent temperature distributions within the upper crust are also suitable as prospecting methods for thermal reservoirs. These prospecting methods are supplemented by a large group of qualitative methods, including other geochemical, geological as well as geophysical methods suitable for locating and mapping thermal water discharge paths.

94

6.1.1 Geochemical and Geological Methods

The prospecting methods dealt with in this section can be used first to localize thermal water discharge paths. This is possible because when volcanic activity occurs, acidic thermal water changes the surrounding rock by alteration. Secondly, with the use of the SiO_2-thermometer (see Sect. 5.1.1.1) quantitative information about the heat source can be obtained. Organic inclusions in sedimentary rock can provide clues as to the geological history and type of thermal reservoir.

6.1.1.1 Mapping Hydrothermal Alteration

This method of prospecting is employed in young volcanic regions where scattered volcanic activity accompanied by steam vents and fumaroles indicates the presence of cooling magma chambers. Visible features on the surface such as steam, water and gas are short-lived. Whereas paths of ascending fluid are closed, new cracks are formed in other weaker zones. This may be due to tectonic activity. However, the old discharge paths remain evident even after the magma chamber cools off. On the surface one can map the region of hydrothermally altered rock although this area is often covered with vegetation and therefore is not fully visible [6.47, 6.51].

After determining the location of a thermal anomaly the next problem is to place it in geological time. For this purpose the age of hydrothermal deposits (sinter, sulphur, limonite etc.) must be determined. Also the relative age of the geothermal anomaly can be obtained by observing sediments within and outside the alteration zone. For example, conglomerates which include thermally altered rock fragments are younger than the hydrothermal event. The age of this altered rock provides the upper age limit.

The duration of hydrothermal activity and the volume of altered rock yield a rough estimation of strength of the heat source.

Results of such investigations at Nigorikawa in North Japan, which is a relatively small thermal anomaly, have been summarized in [6.51]. This anomaly is part of a young Pleistocene caldera having a diameter of about 2 km. In this area there are 28 hot springs (43–92 °C) as well as 7 fumaroles (16–18 °C). The hydrothermally altered zone is located with the fumaroles at the north-east edge of the caldera and encompasses about 0.7 km². In the center of this area are found α-christobalite, amorphous silica and alunite. The outer zones are argillized. Volcanic rocks in which the alteration zone is located are 2.1 million years old. However, these rocks are covered with an unaltered young Pleistocene layer, about 11,500 years old. Thus, it can be assumed that the main phase of hydrothermal activity occurred more than 11,000 years ago.

6.1.1.2 Investigations of Thermal Water

The chemical composition of thermal water provides information concerning the origin of the water and often the temperature of a subsurface reservoir. Water chemistry can be the key to understanding water migration throughout the entire hydrothermal system. In particular one can determine the partitioning between magmatic and meteoric waters as well as fossil pore water from sedimentary rock. An understanding of mixing processes gives important information for effective use of thermal reservoirs. If the water originates from a confined and finite reservoir one must estimate its storage capacity. For a system where meteoric water penetrates to a great depth, is heated and returns to the surface one must estimate the amount of heat which can be removed such that this continuous system remains undisturbed. The difference in both systems can only be determined by thermal water analysis because estimations based on temperatures in the subsurface are in both cases the same [6.9].

Besides determining the structure of the hydrothermal system, the SiO_2 content of water can be used to estimate the temperature of subsurface hot water reservoirs (Sect. 5.1.1.1, [6.47]) as well as quantitative information as to the ratio of surface water to thermal water.

An interesting and well-documented hydrothermal system is in the Hakone volcano field (Fujiyama group) [6.41] of Japan. Magmatic, salt and meteoric water mix in the subsurface. The principle volcano formed 0.4 million years ago, but lava was produced as recently as 5000 years ago. Since the occurrence of a steam explosion 4000 years ago this region has been characterized by expansive fumarolic activity. Thermal water is obtained from about 300 boreholes having an average depth of 500 m. The chemical composition of thermal water allows the discharge area to be divided into four zones based mainly on the anion content (see Table 6.1).

Zone I   includes acid-sulphate water in the vicinity of the fumaroles.
Zone II   includes waters of moderate temperature which have a high bicarbonate and sulphate content. The high HCO_3^--content can be traced back to thermally induced reactions of fossil plants intercalated in volcanic sediment (see Sect. 6.1.1.4).
Zone III is made up of high chloride ($NaCl$-) water whose origin can be found in the central part of the thermal anomaly in the fumarole region. Water migrates within an aquifer in the direction of the pressure decrease and reaches the surface on the slope of the Hayakawa valley where the aquifer is intersected.
Zone IV   the last and largest region is characterized by mixed water as shown in Fig. 6.1 using a three end-member diagram.

Mapping of the so-called thermal water zones together with their interpretation makes it possible to model this hydrothermal system (Fig. 6.2). A groundwater flow system exists in which water flows in a west-east direction and in the central region mixes with discharging volcanic fluids, e.g. water and steam.

Table 6.1. Anion content [ppm] in water of the region of Hakone/Japan, based on [6.41]

Zone		Cl^-	HSO_4^-	SO_4^{2-}	HCO_3^-	$CO_3^{2-} + CO_2$
I	Acid-sulphate water	7	52	526		
II	Bicarbonate-sulfate water	20		381	590	16
III	NaCl-water	2568		82	30	2
IV	mixed water, 2 samples	617/		226/	5/	
		549		85	0	

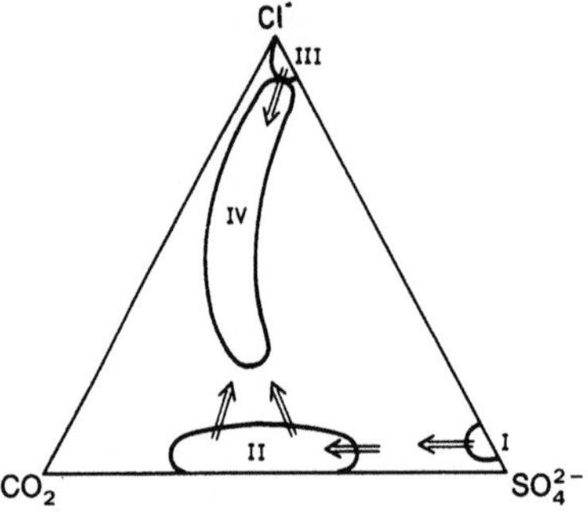

Fig. 6.1. Chemical composition of surface water of various zones in the Hakone region/Japan. This is demonstrated with the $CO_2 - Cl^- - SO_4^{2-}$-diagram showing the process of mixing after [6.41]

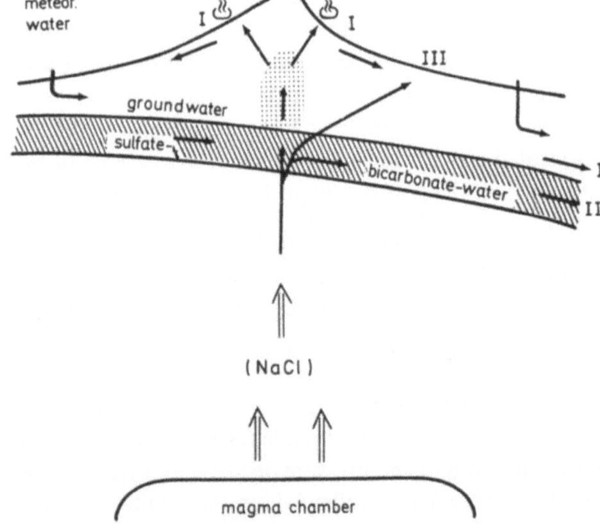

Fig. 6.2. Schematic representation for the origin of various surface waters in the Hakone region of Japan (as in [6.41])

In this central region, at a depth of a few km the temperature and steam pressure are high enough to dissolve large amounts of salt. After mixing with groundwater these thermal waters reach the surface at various places within zone III.

Prior to lowering of an exploitation borehole one can estimate the chemical composition of thermal water reservoirs on the basis of assumed mixing process of the various waters involved.

Also by surveying the temperature distribution in the area [6.23] one can estimate the water temperature to justify the economical prospects.

6.1.1.3 Trace Elements in the Soil

Geothermal anomalies are not always marked by thermal water activity at the surface. Fumaroles or thermally related alteration zones may not be present. In such a case the soil content of radon [6.29], carbon dioxide as well as volatile elements such as mercury, arsen and boron [6.28] can be used for geochemical prospecting of high temperature underground regions. The distribution of above mentioned chemicals is also useful for mapping fault zones and fractures and for determining the extent of deep-lying thermal sources.

The inert gas radon, with a half-life of 3.8 days, is a short-lived element (^{222}Rn). Radon is formed when radium (^{226}Ra) decays and is evenly distributed within the upper crust. Radon rises to the surface in areas of good vertical permeability and therefore mapping radon content in soil can be used to trace fault zones. Because of the short half-life, radon gives important information regarding the transport properties of the subsurface region. This includes rock permeability and a flow velocity of the porefilling fluids within the thermal reservoir [6.29].

In South Japan (Kyushu) many thermal reservoirs show elevated boron and mercury content in the thermal water. Occassionally higher concentrations of arsenic,

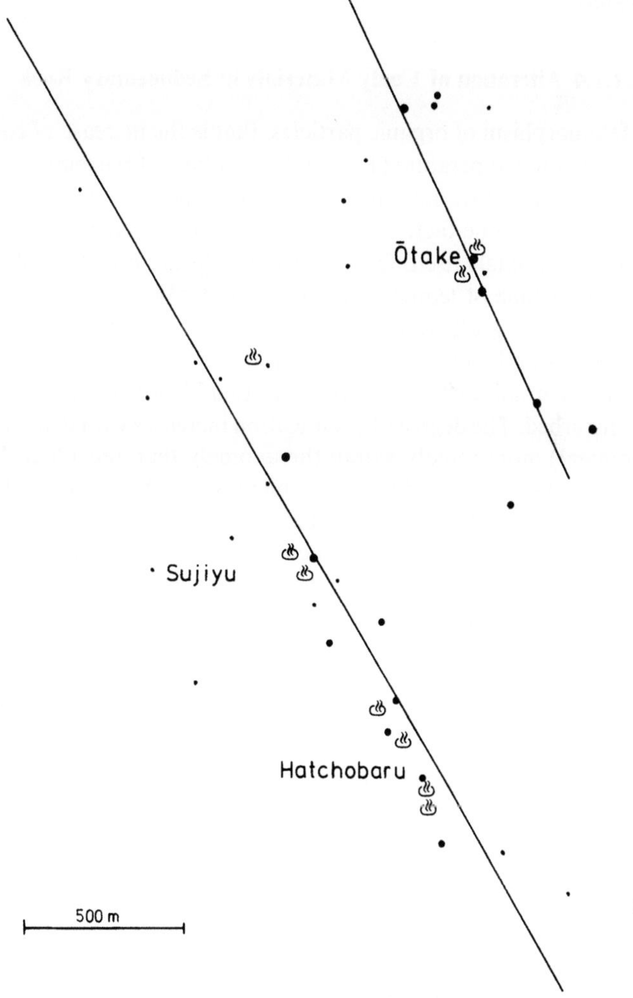

Fig. 6.3. Mercury content ($\bullet, \bullet, \cdot$) in the vicinity of geothermal anomalies of Otaka and Hatchobaru (South Japan) after [6.28]. Positions of fault zones and thermal springs are also shown

98

ammonia and fluorine content are observed. Mapping of the soil mercury content (Fig. 6.3) in the geothermal areas of Otake and Hatchobaru shows large fault zones very clearly in which mineralized water rises. The type of mineralization often leads to the recognition of magmatic fluids although their part in a geothermal system is minor. By far the greatest part of the thermal water systems consists of surface water. In one case, surface water may be stored in pore spaces of sedimentary rock and convects towards the surface when temperatures are increased. In another case, as in areas characterized by large scale fracture tectonics, surface water may form a water circulation system as cold meteoric water migrates down to greater depth, heats up and rises to the surface in another place. During the heating of water at depth a solution equilibrium with the rock matrix can be obtained in which salts or ore minerals are dissolved so thermal water appears magmatic. For this reason there are no reliable geochemical thermometers based on contents of trace elements in thermal water.

6.1.1.4 Alteration of Coaly Materials in Sedimentary Rock

Metamorphism of organic particles, that is the increase of coalification, is controlled by reaction temperature (T > 35 °C) and time of the temperature exposure. As degree of coalification within a thermal anomaly provides information about an integrated temperature-time factor (see Sect. 5.1.5), it therefore is useful in characterizing the thermal anomaly itself. Quantitative results are possible only in rare cases because of the short time of temperature exposure [6.11].

Regional geothermal conditions over long periods of time control coalification processes and form a definite pattern for various ages of coal at different depths. When thermal water or magma rises to a higher crustal horizon this regional picture is disturbed. The degree of coalification increases with depth (i.e. vertical coalification gradient) more rapidly within the anomaly than outside it. Within the Cerro Prieto geothermal field in Mexico, an increase in the optical reflectivity of the organic particles of 2.8 %/km is reported [6.6], which exceeds a few times most common gradients. After prolonged influence a temperature gradient can be associated with a coalification gradient [6.8]. If the heating time is short this cannot be done without additional thermal model calculations, e.g. cooling of magma, because the progress of coal metamorphism cannot reach values during a relatively rapid cooling of an intrusive body which would be reached for comparable reaction temperatures under steady state conditions. Nevertheless higher local coalification gradients can prove whether there is or has been a thermal perturbation or not. For example, the higher coalification gradients of the middle Upper Rhinegraben reflect a higher temperature gradient than in the neighboring Molasse trough. Overprinting this generally elevated heat flow density are secondary effects, which can best be observed in the Landau/Pfalz region [6.54]. Here coalification gradients within the many boreholes deviate sharply from a high average value [6.52]. Undoubtedly the strong local deviations indicate a convective perturbation to the heat flow density. For the middle Upper Rhinegraben it can be shown that the present regional thermal anomaly has existed only for a short time (see Sect. 5.1.5).

The occassional but quite considerable changes in degree of coalification within a small area can be traced back to the oxidizing influence of thermal water which migrates along cracks in the rock. In borehole Urach 3, situated in the Swabian thermal anomaly, influence of thermal water circulation is observed in the Rotliegend and Stephanich layers [6.10].

The alteration of organic substance with temperature increase can also be traced through one of its reaction product, natural gas. Regions of highest temperature can be marked by looking at its concentration.

6.1.2 Geophysical Methods

Geophysical prospecting methods should be suitable for localizing deep thermal reservoirs by measurements performed at the surface. Direct methods in thermal prospecting measure temperature or heat loss directly. Examples include the measurement of soil temperature from the air through infrared techniques or temperatures in the shallow subsurface by means of a thermometer. Once temperatures have been determined at various depths below the surface a heat flow density may be deduced which is better suited to estimating the strength of a heat source. Indirect methods of thermal prospecting utilizes the temperature sensitivity of physical rock properties to make quantitative estimates of temperatures (compare Sect. 5.2.2). Qualitative methods of prospecting such as mapping seismicity and electric self-potential at the surface are also employed to locate thermal reservoirs.

6.1.2.1 Infrared Measurements

The thermal infrared region includes wave lengths from 3 to 14 μm. The two atmospheric windows which lie between 3 and 5 μm and 8 and 14 μm respectively are within this region. Longer IR waves become absorbed by the atmosphere. If a black body radiates maximum energy at wave lengths of $\lambda = 10$ μm its temperature will be

$$T[°K] = 2.9 \cdot 10^3/\lambda_{max}[μm] = 290 °K. \tag{6.1}$$

The radiation within a given IR band cannot be photographed with an ordinary camera because the glass lenses are opaque to IR radiation. Thus there are special IR cameras [6.51] such as those which use semiconductor detectors. For surface measurements radiometers are used, which measure the radiation temperatures by means of a photocell displaying a resolution of $\Delta T \leq 0.2$ °C. Scanner-systems based on thermal imaging are developed with a much higher resolution of 0.01 °K [6.46, 6.50] from airplanes as well as satellites. Sensitive Scanner systems are most effective in airborne prospecting. Measured radiation temperatures depend not only on solar radiation but also on topography of the region and on thermal properties of the soil. Thus, soil albedo (see Sect. 4.1.1) plays an important role as well as thermal conductivity and thermal diffusivity. The temperature contrast which is caused by different soil thermal properties is greatest at night just before sunrise and also in daytime about an hour after the mid-day temperature maximum [6.46].

Thermal energy flux through the surface consists of three parts:

$$Q = Q_S - Q_R - Q_A \tag{6.2}$$

Q_S from the incident solar flux which becomes absorbed during the day, Q_R reradiated by the surface and an anomalous part Q_A related to a temperature anomaly. The magnitude Q_S is generally so large that Q_A cannot be calculated with confidence. Only at night when $Q_S = 0$ can Q_A be determined and only so long as the temperature anomaly ΔT_A is large enough. At night the radiation temperature T_S is given by:

$$T_S = (T_0 + T_A + (Q_0 \sqrt{\kappa/\omega}/K) \cos \omega t)\, \varepsilon^{0.25} \qquad (6.3)$$

where a maximum heat flow density Q_0 is generated by the sun at time $t = 0$ (at noon). κ and K are the thermal diffusivity and thermal conductivity of the soil, ω the angular frequency of the daily solar radiation, ε emissivity of the soil and T_0 is the average soil temperature.

In order to recognize a thermal anomaly through IR-measurements, the anomalous heat flow density from underground must be 100 times greater than normal [6.26]. This occurs only for very strong, mostly volcanic-related anomalies. IR-methods can be applied from an airplane or satellites in order to recognize thermal springs on land and in water. For very large anomalies which reach more than a thousand times the background heat flow density, an empirical equation [6.48] permits one to calculate heat flow density from surface temperature. In summary, IR-methods may be useful as a reconaissance method in some geologic terrains, but generally temperature measurements taken at the earth surface with a thermometer are more accurate and quicker in localizing thermal sources.

6.1.2.2 Shallow Temperature Measurements and Heat Flow Density

Measurement of surface and near-surface temperatures can also be included as a geothermal prospecting method. If temperature distributions are measured at two depths, under favorable circumstances one can estimate the strength of the heat source. The temperature field can also be used to determine sites for exploitation drillholes. The areal mapping of temperatures must satisfy two conditions. First, it must be independent of daily temperature variations. This can be accomplished if the measurement depth is greater than 0.5 m (see Sect. 5.2.1). It is common to measure temperatures at depths of $z = 1$ m and $z = 10$ m [e.g. 6.46]. Second, the measurements should advance quickly because these depths are still influenced by annual variations in the surface temperature. If the survey is completed in a short time, then effects of the long period annual wave are small.

Electric measuring devices such as resistance thermometers (see Sect. 5.2.1), thermistors and thermocouples are commonly used as thermometers. With a measuring accuracy of $\pm\, 0.1\, ^\circ$C and a depth interval of about 10 m, the average temperature gradient of $0.30\, ^\circ$C/m can only be obtained if a correction for the annual temperature cycle is made. In contrast to background conditions a shallow depth economical thermal reservoir produces temperature gradients above it of about $0.1\, ^\circ$C/m. In this case prospecting will succeed by measuring temperatures at two depths with a 10 m difference. Any estimation of the heat stored in the reservoir will be very rough if it is based on these shallow measurements. Only using drill holes of about 30 m or deeper, below the influence of the annual surface temperature cycle, can one make more quantitative calculations.

Near-surface temperature variations are very effective in mapping zones of convective heat transport within a thermal area [e.g. 6.25, 6.26, 6.46]. In particular fault

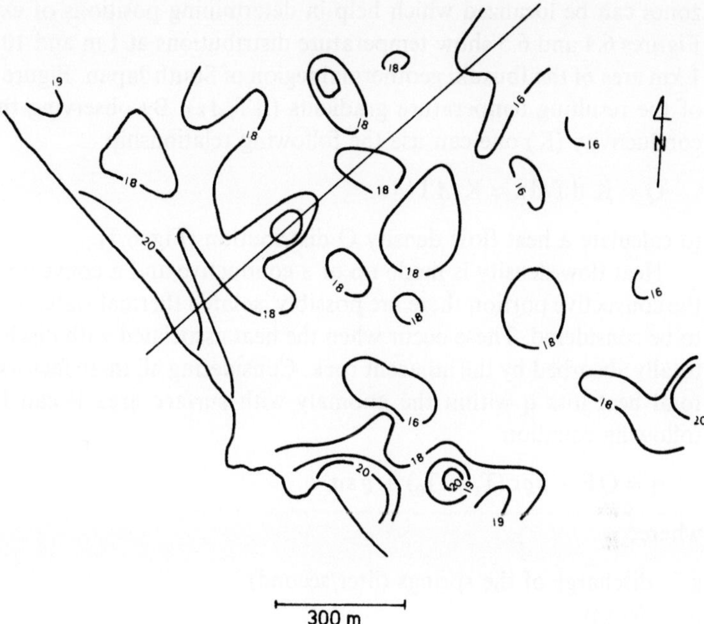

Fig. 6.4. Temperature distribution [°C] at 1 m depth in the geothermal anomaly at Ibusuki (South Japan) (after [6.56])

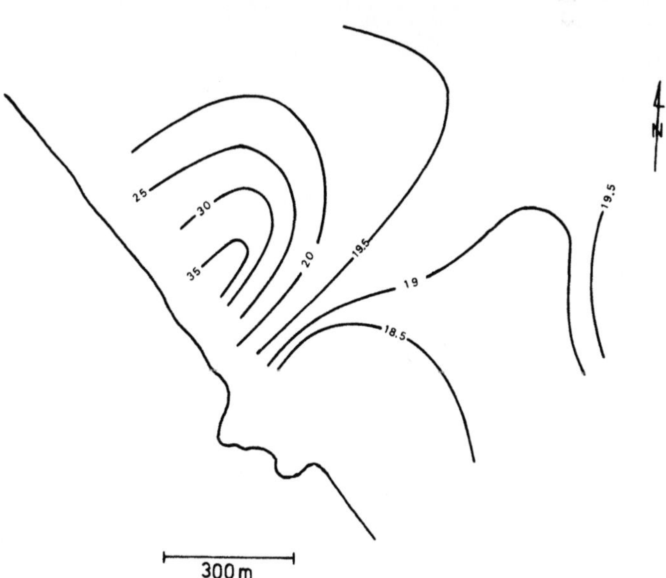

Fig. 6.5. Temperature distribution at 10 m depth [°C] in the geothermal anomaly at Ibusuki (South Japan) (after [6.56])

102

zones can be localized which help in determining positions of exploitation drilling.
Figures 6.4 and 6.5 show temperature distributions at 1 m and 10 m depths within a
1 km area of the Ibusuki geothermal region of South Japan. Figure 6.6 shows a profile
of the resulting temperature gradients ($\Delta T/\Delta z$). By observing the varying thermal
conductivity (K) one can use the following relationship

$$Q = K \, dT/dz \approx K \, \Delta T/\Delta z \tag{6.4}$$

to calculate a heat flow density Q distribution (Fig. 6.7).

Heat flow density is made up of a conductive and a convective part. Along with
the convective portion there are possibly isolated thermal water and steam discharges
to be considered. These occur when the heat associated with discharging fluids is not
totally absorbed by the adjacent rock. Considering all these factors an estimate of the
total heat loss q within the anomaly with surface area F can be made using the
following equation

$$q = QF + s\varrho c(T_1 - T_0) + \varrho s\sigma \tag{6.5}$$

where:

s   discharge of the springs (liter/second)
ϱ density
c   specific heat of the thermal water or steam
σ heat of condensation in the case of steam sources
T_1 temperature of the upper thermal water reservoir
T_0 temperature of the earth surface.

The heat loss q gives an indication whether the thermal anomaly under economic
use would be a depletable deposit or if it has adequate capability for continuous
steady state use.

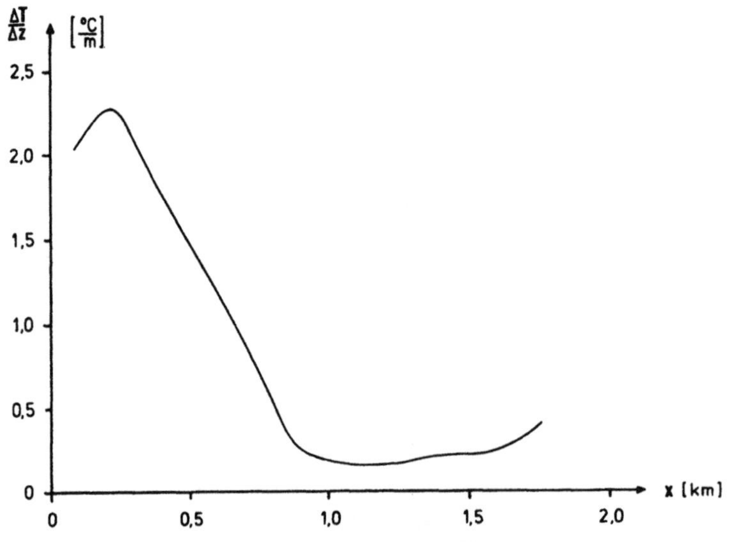

Fig. 6.6. Near surface temperature gradient profile across the geothermal anomaly at Ibusuki
(South Japan). Profile location is shown in Fig. 6.4

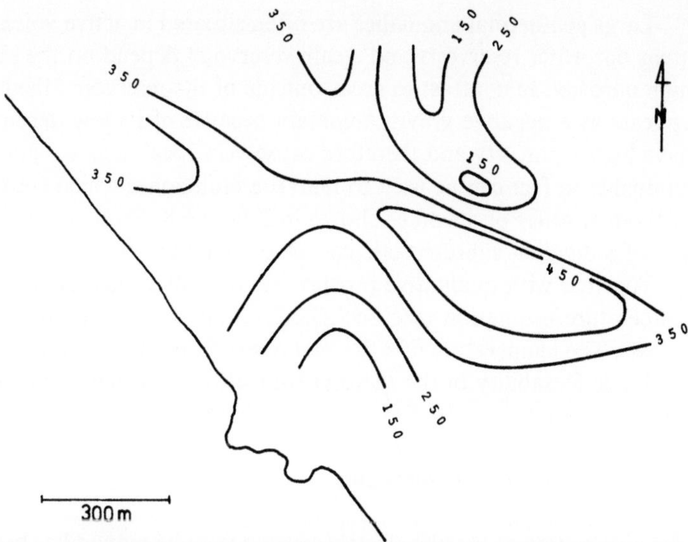

Fig. 6.7. Surface heat flow density [mW/m²] for the geothermal anomaly at Ibusuki (South Japan after [6.56])

6.1.2.3 Gravimetric Measurements

Gravimetry alone is not an adequate prospecting method for unknown thermal reservoirs. However, its application within a thermal anomaly can be very helpful.

Gravimetric results are primarily useful in determining geological structures. These in turn furnish important clues concerning the position of maximum heat within a thermal anomaly and for the location of exploration and exploitation drill holes.

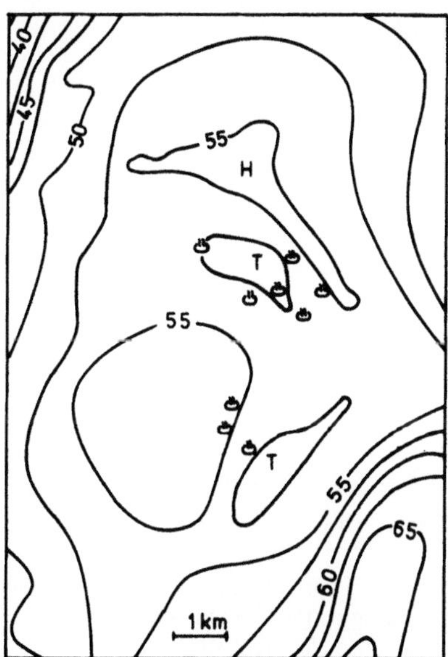

Fig. 6.8. Bouguer gravity map from the region of the geothermal anomaly at Hakone (Japan) [6.26]. Gravity lows are labeled T and highs are labeled H. Thermal springs are also shown

Large geothermal anomalies are often situated in active volcanic regions. In such areas hot water reservoirs and steam reservoirs depend on the existence of layers of high porosity. In contrast to areas outside of the reservoirs, the high porosity region appears as a negative gravity anomaly because of its low density. Fault zones also have higher porosity and therefore cause very local negative gravity anomalies. The anomalies in fact can be used to map the fault zones which contain discharge paths of thermal water or steam as shown in Figure 6.8. From a system of fault zones the size of individual subreservoirs can be estimated.

Together with qualitative prospecting it is often possible to make a quantitative temperature estimation (see Sect. 5.2.2.1) based on thermally induced density decreases. The temperature of a thermal reservoir is an important factor in determining economic feasability of the thermal source for a particular energy use.

6.1.2.4 Geoelectric Measurements

The distribution of specific electric resistance underground has been used successfully in groundwater prospecting for many years. With increasing interest in geothermics, geoelectrics has played an equally important role in prospecting for thermal reservoirs. Electrical methods can be divided into two groups. First, there are several methods which utilize artificially induced electric fields. Second, are those which

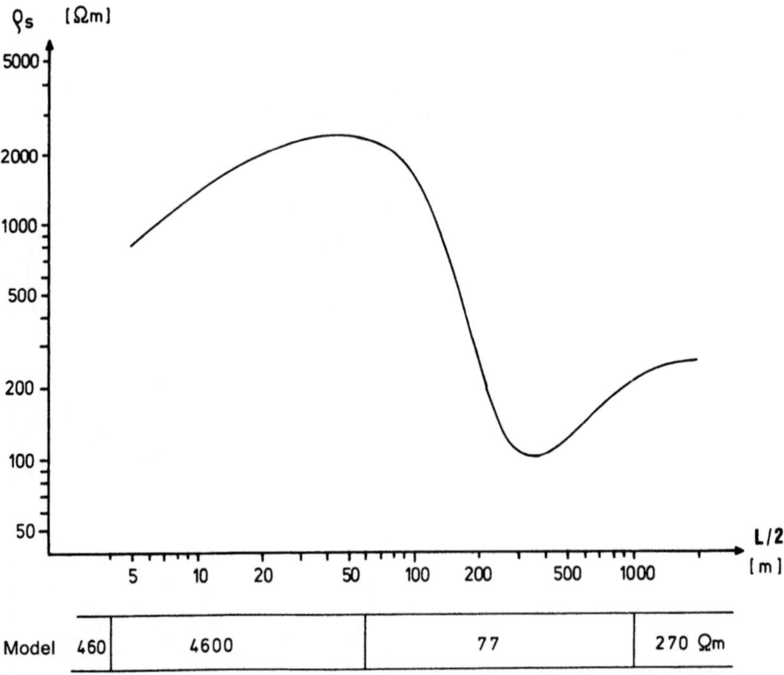

Fig. 6.9. Top: Apparent specific resistance (ϱ_a) versus electrode spacing (L) for a Schlumberger sounding in the thermal anomaly of Otake (South Japan) after [6.42]. Bottom: Model of the subsurface electrical structure

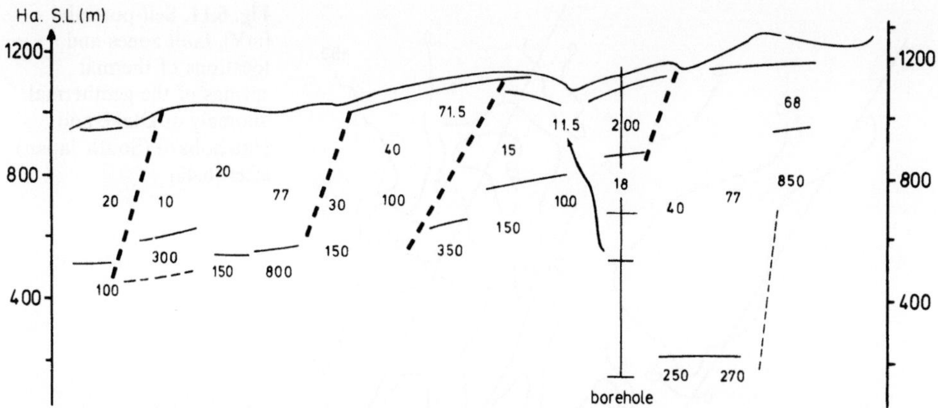

Fig. 6.10. Distribution of specific resistance (Ω m) in a cross section of the geothermal anomaly at Otake (South Japan) after [6.42]

involve the measurement of the electrical self-potential of naturally occurring electric fields.

Geoelectric techniques are used for vertical sounding as well as for lateral mapping of layers of high conductivity. Layers of low specific resistance can generally be identified with thermal water reservoirs (Fig. 6.9). A sudden decrease of specific resistance identifies the aquifer [e.g. 6.47].

Specific resistance is controlled by porosity, salt content of the water and temperature (see Sect. 5.2.2.2). In geothermal reservoirs of South Japan, thermal water is stored in fractured hydrothermally altered andesite at a few hundred meters depth. There the specific resistance is effectively proportional to the degree of weathering, which determines the average porosity of the layer. Because temperature and salt content can be equated in these geothermal systems, for these specific reservoirs it is possible to obtain an estimation of individual reservoir potential from their specific resistance distribution [6.42]. From various sounding curves a subsurface cross section can be constructed which illustrates the nature of fault zones. Their pattern (Fig. 6.10) thus helps to determine the location of exploitation boreholes.

In the case of natural electric fields, one measures a self-potential which is generated when electrolytes migrate through a permeable layer (= streaming potential). These electro-kinetic potentials considerably exceed thermo-electric potentials which also occur in a geothermal area [6.14, 6.39]. Because a very active hydrothermal system often exists within a geothermal area, the self-potential method can be well suited for prospecting of thermal reservoirs. A calculation of filtration potential is difficult because of the numerous parameters to be considered such as pressure difference, viscosity and specific resistance. Generally potential difference increases with pressure difference as well as with specific resistance of pore fluids. The polarity of the potential-drop is generally negative for descending water and positive for ascending water [6.35].

The thermo-electric effect occurs wherever there is a temperature difference in a composite medium. In this case the electric potential-drop (ΔU) is proportional to temperature difference. The proportionality factor is within the range of $\Delta U/\Delta T \approx 0$ and about 1.5 mV/°C [6.14]. A quantitative evaluation of self-potential measurements

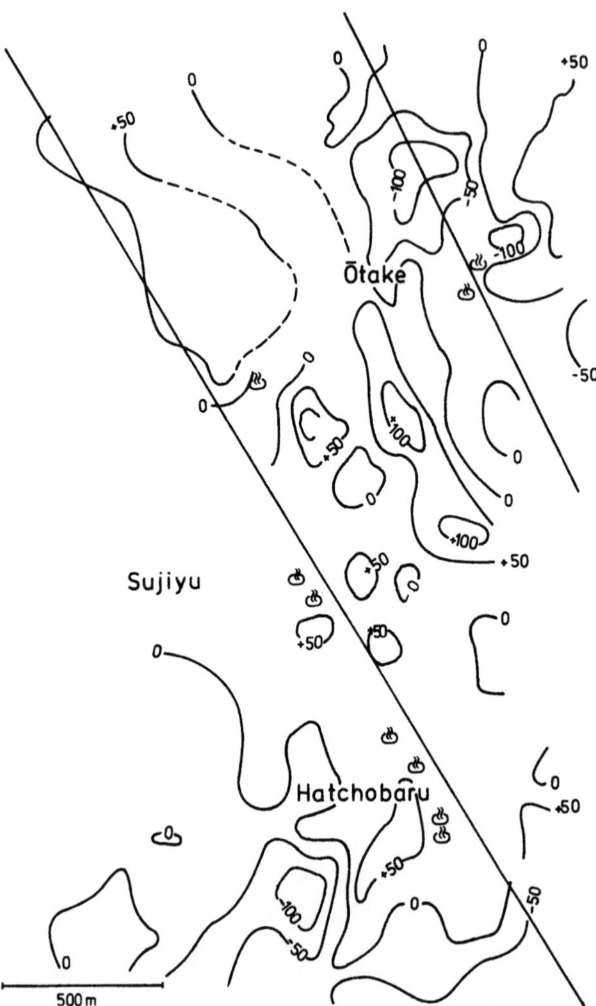

Fig. 6.11. Self-potential [mV], fault zones and locations of thermal springs of the geothermal anomaly at Otake and Hatchobaru (South Japan) after [6.42]

with respect to temperature estimation or to quantity of discharging thermal water is not possible at this time. This is due in part to difficulty with the theory of self-potential but also to the superposition of disturbances to the self-potential such as telluric currents, topographic streaming potential, electrochemical effects and others.

An example of self-potential measurements is given within the geothermal anomaly at Otake/Hatchobaru, South Japan [6.42]. This thermal anomaly is associated with large positive and negative self-potential effects but are more or less enclosed by a 0 mV equipotential contour (Fig. 6.11). The map also shows a good correspondence between self-potential anomalies and the position of thermal springs and steam discharges.

6.1.2.5 Seismic Methods

Reflection and refraction seismology methods have been used only in a few regions for thermal reservoir prospecting [e.g. 6.33]. Travel-time residuals of compressional

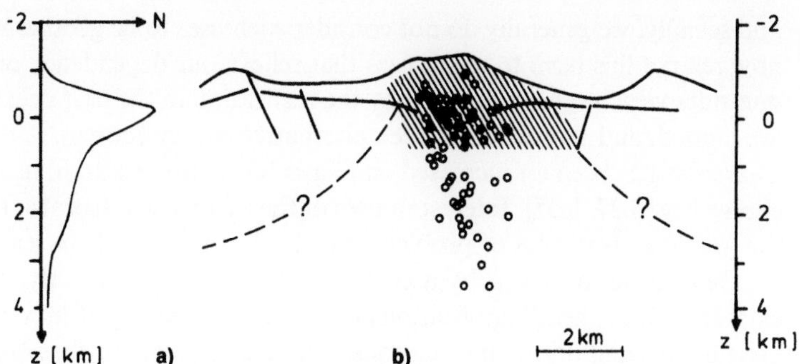

Fig. 6.12 a, b. Distribution of micro-earthquake hypocenters in the vicinity of Hakone/Japan and temperature range over T = 100 °C (shadowed area) as in [6.41]
a Frequency-depth distribution
b Distribution in the upper crust (W-E profile)

waves are generally negative if the waves have passed through a geothermal anomaly. The delay in the travel time is about 0.2 seconds in areas examined. This can be traced back to two characteristics of hydrothermal heat reservoirs – increased porosity and higher temperatures. In addition to the velocity decrease of compressional waves within the anomaly, the amplitude and wave form are also altered [6.33]. The absorption coefficient is larger within a thermal anomaly (see Sect. 5.2.2.5). Reflection and refraction seismology can be better employed for locating deep-lying thermal sources than in prospecting for shallow reservoirs.

Steam or water rising within a geothermal anomaly cause high local temperature gradients along their ascent paths. If these lead to faulting in the rock, then sudden pressure release may cause local steam explosions of super-heated water. Such local events within a geothermal anomaly are observed as an increase in seismicity i.e. micro-earthquakes [e.g. 6.24, 6.53]. It is therefore possible to localize thermal reservoirs with temperatures far above T = 100 °C, the boiling point of water under normal pressure. An example from mid Japan is shown in Fig. 6.12. In the Hakone region, belonging to the Fujiyama volcano group, micro-earthquake centers lie between the surface and a depth of about 5 km. The epicenters cover an area of about 7 km. This area coincides with the maximum geothermal anomaly, thus giving an example of the connection between thermal reservoirs and increased seismicity.

6.2 Utilization of Geothermal Energy

The oldest form of geothermal energy utilization by people may have been the uses of thermal water in bathing pools. At the time of the Roman Empire thermal baths were highly developed in Europe and along the eastern Mediteranean. The main interests at that time were therapeutic and social not economical; another form of using geothermal heat as an energy source. In latitudes of colder climate, as in Iceland where geothermal energy was plentiful, further possible uses for hot springs were found (e.g. cooking and heating). Even though today spas are still valued medicinally and socially, we generally do not consider such uses to be geothermal utilization. We now reserve this term to mean uses that relieve our dependency on fossil fuels. This one-sides view has been fostered by the realization in the past decade that fossil fuels are limited, and that we must seek alternative energy sources. A direct result of this awareness has been an increased emphasis on energy research, including geothermal energy [e.g. 6.27, 6.55]. Interest in uses of the earth's heat has also been intensified by the growing awareness of problems involved in the use of nuclear energy.

Geothermal heat can be used for heating purposes directly. Many possibilities exist for "direct heat" applications including the heating of homes, public facilities, greenhouses and drying installations. However, in spite of the simplicity and attractiveness of these uses the present cost of installation is so high that it is economical only in limited regions where thermal water transport is not costly.

Thus, the utilization of geothermal heat of low enthalpy has been somewhat ignored in terms of energy research. At present the main goals of geothermal research are concerned with locating higher enthalpy sources and converting thermal energy to easily transportable electric energy since this form of energy has much broader use. Conversion of geothermal heat to electricity can be accomplished in two ways. First, steam from drillholes can be used directly to drive the turbine of a generator. This method is in general use today. In the second case, if thermal energy is stored in water just below its boiling point, then the energy must be transferred to a secondary fluid in a closed system by means of a heat exchanger. If the secondary fluid has a boiling point lower than that of the thermal water then its vapor can be used to drive a turbine. At present there are only pilot systems for this second method.

Hot water and/or steam transported from underground contains various amounts of dissolved minerals, causing not only technical but also environmental problems. These are fundamental obstacles in geothermal energy utilization.

6.2.1 Use of Thermal Water in Swimming Pools

Healing properties of certain springs were known in ancient cultures dating back, for example, to the 13th century B.C. in ancient India. Evidence of prehistoric use of well known springs such as Bad Pyrmont, Germany has been found at the bottom of springs in the form of cult items.

Thermal springs acquired intensive use and renown in ancient Greek and Roman times. Asklepiades is recognized as first giving therapeutic springs an accepted role in medicine [6.7]. Later the Romans built bathhouses which eventually developed into complex installations with hot pools, sunbaths and physical training halls. Claudius Galenos was a great promoter of thermal pools during the second century A.D. Under Caesar M. Aurelius Antonius this bathing cult reached its peak and extended throughout Europe. Noteworthy grand public baths in Germany dating from Roman time are Aachen (Aquae Grani), Baden-Baden (Aquae Aurelia) and Wiesbaden (Aquae Matiacae). From a historical perspective the use of these public as well as private pools constituted an early peak in the use of geothermal energy.

Bathing cults also developed in other regions of the world where hot pools were available. In many localities, such as Japan, they still play an important role. For example, hotels outside of large cities in Japan often have their own thermal pools.

These attract guests by their therapeutic reputation, too. Many thermal pools possess a number of trace elements which are known to have healing qualities. Also, areas with thermal pools are tectonically young with spectacular scenery; so guests are attracted to these regions for recreation possibilities.

These therapeutic pools are almost always related to residual volcanic activity. Hot springs, steaming ground, mud pots and fumaroles also mark these areas. Their worldwide distribution is heavily biased to the boundaries of lithospheric plates either on ocean ridges (e.g. Iceland) or volcanic arcs at convergent plate boundaries (e.g. Japan).

Another class of thermal springs is of sedimentary origin. Hydrothermal systems consisting of pore water as well as meteoric water form during subsidence of a sedimentary basin. In places of comparatively low vertical permeability vadose water is warmed at depth and rises to form thermal springs (e.g. Pannonian Basin of Hungary). In these systems waters of different origin commonly mix in the subsurface, making a meaningful classification based on water chemistry difficult [e.g. 6.23].

The utilization of thermal water in Hungary as in other regions in Europe was promoted by the Romans. Even today many of these pools, encountered nearly 2000 years ago, are still in use. In recent years the number of pools have increased because unsuccessful petroleum exploration wells are used for obtaining thermal water. In Hungary 343 thermal springs are in use [6.6], of which more than half supply water to swimming and therapeutic pools.

A further use of mineral springs is for the sale of bottled mineral water. In Europe, where mineral water consumption is considerable, this use of thermal springs is a significant contribution to the local economy.

Although the use of thermal water in pools is only a small aspect of geothermal energy, it does illustrate that this energy form has a far broader scope than simply as an alternative to oil. Thermal springs contribute directly or indirectly to restoring human energy. This use of geothermal energy cannot be expressed in megawatts nor is it included in statistics of energy consumption.

6.2.2 Thermal Water for Space Heating

In regions of moderate and colder climates a large part of the annual energy consumption is for heating purposes. The required heat is obtained nearly entirely from fossil fuels. Since relatively low temperatures are necessary for heating, geothermal heat would seem to be ideally suited for this purpose. Indeed many people maintain that the future of geothermal utilization is in this area of space heating. At present the principal obstacle to large scale use of geothermal heating is not technical but financial. The investment necessary, mostly in borehole costs, is still too high for economic replacement of fossil fuels with thermal heat. However, if the cost of coal, gas and oil increases, time will come when the low operating cost will compensate for the high installation cost.

Geothermal heat is available to some extent everywhere on earth. However, because temperatures generally increase slowly with increasing depth and because of the high cost of drilling, it is only economical to use this energy in a few favorable locations. In Germany these are the Upper Rhinegraben region and the geothermal anomaly near Urach in the Swabian Alb (see Sect. 4.1.5). In Northern Germany temperature gradients above salt domes are higher than the regional average value [e.g. 6.1.1] because of thermal conductivity differences between salt domes and the adjacent sediment.

Thermal reservoirs of quite low temperature can now be exploited using heat pumps. Heat pumps operate by using a low boiling point fluid in a closed system. The fluid is evaporated at one place by drawing heat of vaporization from its surrounding. The vapor is then compressed and heat of condensation released at another place. In this way heat can be extracted from the earth and released in homes and other building. Such pumps can be used even at the low soil temperatures controlled by our climatic conditions. The disadvantage of this system is that it requires a large surface area from which it can extract heat. For example, the area of an entire garden is necessary to supplement the heating of a one family home.

However the potential use of heat pumps is much more extensive than simply in single house space heating [e.g. 6.2]. Water at low temperature can be brought to a higher temperature in order to connect into an existing heating system. Also several heat pumps can be connected in series to produce a cascade effect.

Extraction of heat from groundwater is convenient as well as technically simple, but the use of groundwater in many cases causes a disturbance to the flowsystem. Thus in populated areas, groundwater aquifers cannot be considered as thermal reservoirs for heat pumps. It seems a preferable practice to withdraw the geothermal energy in the form of deep-seated thermal water from suitable drill holes.

To be economical, thermal water should have a temperature of at least 40–50 °C. If it cannot be used directly because of high mineral content then its heat must be transferred to a secondary system using heat exchangers. In order to prevent environmental damage the cooled water should be reinjected into the aquifer at another place (see Sect. 6.2.4). Petrophysical conditions of the aquifer determine the distance between extraction and injection locations and in turn the lifetime of the thermal reservoir [e.g. 6.19].

Natural hydrothermal systems, such as are used for producing electricity, should supply constant energy over relatively unlimited time. In these systems meteoric water receives continual heating from a magma chamber. It rises to a higher level and is drawn from an aquifer as thermal water. Natural migration of meteoric water insures replacement. For areas without natural hydrothermal reservoirs a new technique has been developed to extract heat from the dry but hot rock [6.20]. Two boreholes in close proximity are drilled into the hot dry region. Artificial cracks are then induced in the region surrounding the boreholes by a process called hydrofracturing. When surface water is pumped under high pressure into one hole it migrates towards the other and thereby extracts heat from its surrounding. The temperature increase of the migrating water is higher when the migration velocity through fractures is slow and when the temperature of the exchange surface is high. At very high temperature water may even flash into steam. Research into this hot dry rock technology takes place mainly in the United States on the volcanically youthful Jemez Plateau, New Mexico [6.49]. Here the financial and technical commitment is rather low compared to the energy obtained [6.31, 6.49].

Depending on the amount of available thermal water, its temperature and the local demand for heat, geothermal energy can be used for space heating in homes, for

Fig. 6.13. Utilization of thermal waters in Hungary with a total achievement of 1.17 GW [6.6]

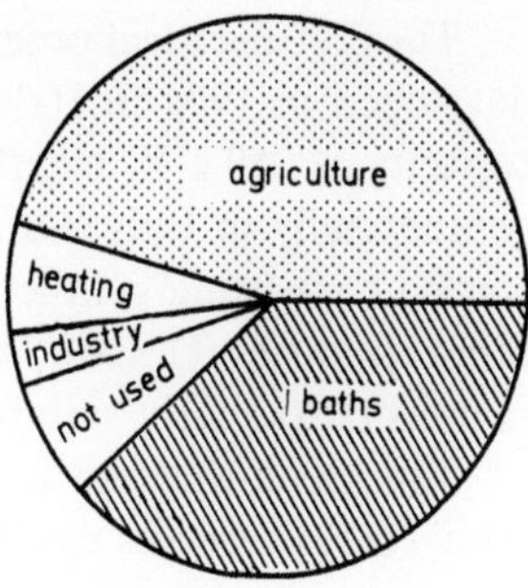

greenhouses, swimming pools and various industrial endeavors. In Iceland and Hungary geothermal space heating is used very intensively. In 1981 in Iceland 70% of all homes were heated geothermally [6.25] and in 1976 geothermal energy made up 31% of all energy consumed [6.18]. For geological reasons geothermal energy applications in middle Europe in the foreseeable future cannot be expected to attain such a large percentage of the energy used as in Iceland. However, it will remain regionally and locally important.

An example of regional uses can be taken from Hungary where the use of geothermal energy is much more advanced [6.6, 6.43]. Here nearly 300 drillholes serve as primary energy sources. Half of these supply water at temperatures exceeding 60 °C. Most of this hot water is used in agricultural enterprises such as greenhouses, animal husbandry and drying installations. Use of lower temperature water in thermal baths and swimming pools constitutes a secondary but still important use as is shown in Fig. 6.13.

Since the mid 1970's all European countries have been making efforts to utilize heat from thermal water in order to lower their dependence on fossil fuels [e.g. 6.4, 6.16, 6.30, 6.47]. Iceland is particularly favored for exploiting geothermal energy. Considerable energy is also available in the Pannonian basin of Hungary although at greater cost of extraction. In France where the average temperature gradient is less than in Hungary, geothermal applications have so far been restricted to the Parisian basin and Rhinegraben [e.g. 6.15, 6.32]. Energy is drawn from thermal water by means of heat exchangers and then fed into an existing heating system. The Federal Republic of Germany has also made an effort to obtain geothermal heat in the Rhinegraben at Buehl from where it will be fed into a heating system [6.47].

6.2.3 Converting Geothermal Energy to Electrical Energy

Beside direct uses of geothermal heat, conversion of geothermal energy to electric energy is of great importance. Electrical energy is notable because of its versatility of service, but even more important because it is easily transportable. It is only necessary to connect into an existing electricity supply grid. A disadvantage of geothermal power-plants in comparison to conventional power-plants is their thermodynamically imposed lower efficiency in converting relatively low temperature steam (200 °C) or water (100 °C) to electricity. Pressures are also considerably lower than in conventional plants.

112

The first electrical generator driven by geothermal energy in the form of steam was developed in 1904 in Italy (Larderello/Toscany). By 1913 this plant already had a capacity of 250 kW. Nearly 200 times as much electric energy is presently produced geothermally in Italy. Other countries followed slowly. In the 1950's New Zealand started converting geothermal energy to electric energy at Wairakei. Ten years later other countries (U.S.A., Japan, Mexico, El Salvador, Iceland and USSR) followed in regions particularly suitable.

The break-through for geothermal energy came only after the realization that fossil energy reserves could soon be exhausted by the world's ever-increasing demand for energy. The annual increase of geothermal power stations is about 20%. These new stations use steam energy early exclusively, converting it to electrical energy.

6.2.3.1 Exploitation of Steam

The economically most promising geothermal systems are those which produce steam at relatively shallow depth. First, steam can be converted to electricity at a higher efficiency than can hot water. Second, the shallow depth is important because production drilling can be accomplished at minimum cost.

Italy, with the longest tradition of geothermal powerplants, produces about ⅓ of the total geothermally produced electricity in the world. The power plants are fed mainly from the Larderello geothermal area, but also from Monte-Amiata. At present they produce about 450 MW electricity. Geothermal power production in the United States began in 1960 with a small 12-MW plant at "The Geysers", California. By 1976, production at this locale had risen to 522 MW; the projected energy supply by 1986 is 1.6 GW [6.2]. New Zealand is the third-ranking country of the world in terms of its geothermal power development. Their first geothermal plant was built in the fifties in a region called Wairakei. Today it produces over 150 MW. In contrast to the exploitation of large geothermal systems, Japan offers an example of exploiting many smaller thermal reservoirs. Here 7 anomalous geothermal regions together produce about 200 MW of power. The advances geothermal energy has made in the past ten years can best be demonstrated by the development in Japan shown in Table 6.2 and Fig. 6.14.

World production of geothermal electricity in 1976 was about 1.4 GW. In 1983 it exceeds 4 GW [6.46]. The annual growthrate is expected to reach 19% as an upper limit and 7% as lower limit [6.46].

Table 6.2. Geothermal power stations in Japan

Location	Year of Commission	Electric Capacity [MW]
Matsukawa	1966	22
Otake	1967	13
Onuma	1973	10
Hatchobaru	1977	50
Kakkonda	1977	50
Onikobe	1977	25
Mori	1982	50

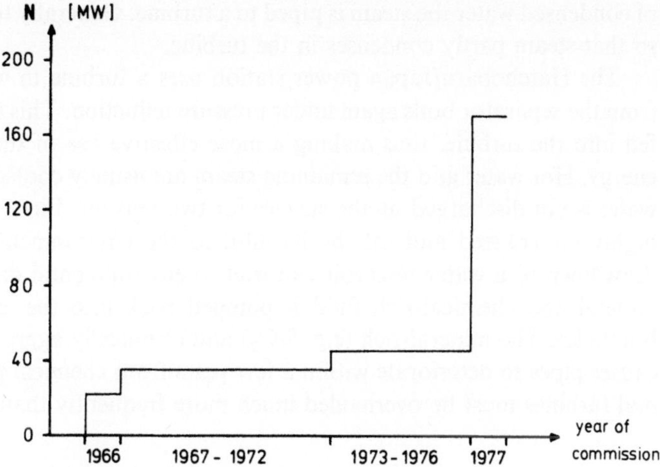

Fig. 6.14. Development of geothermal electricity production in Japan

Methods of operating a geothermal power station are in principle simple. Steam rising in the borehole is fed into a turbine which in turn drives a generator producing electric energy. In practice technical implementation is much more complicated.

Steam rising from the drillhole often contains large amounts of water. The water content depends on the properties of the fluid in the reservoir. It may already be in the form of steam, or it may convert to steam in the borehole as the fluid rises and pressure decreases, thus lowering the boiling temperature. Considerable cooling takes place with vaporization due to the high latent heat of vaporization. At the surface steam and liquid must be separated. After this first step, steam produced from a number of drill holes is collected in one boiler (Fig. 6.15). After a further separation

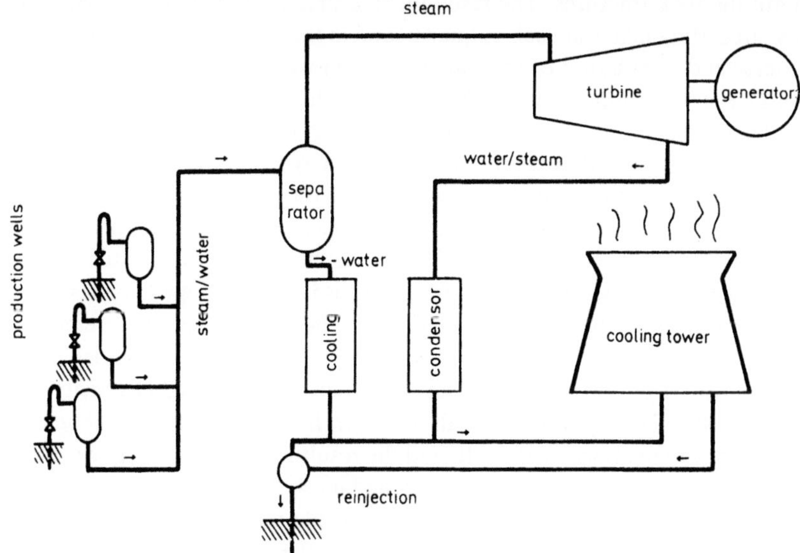

Fig. 6.15. Schematic representation of a geothermal powerstation

of condensed water the steam is piped to a turbine. Generally turbines are constructed so that steam partly condenses in the turbine.

The Hatchobaru/Japan power station uses a turbine in which condensed water from the separator boils again under pressure reduction. This secondary steam is also fed into the turbine, thus making a more effective use of the available geothermal energy. Hot water and the remaining steam are usually cooled to 40 °C. This cooled water is not discharged on the surface for two reasons. First, the water is frequently highly mineralized and can be harmful to the environment (Sect. 6.2.4). Second, drawdown of a water reservoir can lead to environmental damage. As a result this mineral and chemical-rich fluid is pumped back into the earth through injection boreholes. The mineral-rich (e.g. SiO_2) and chemically aggressive geothermal water causes pipes to deteriorate within a few years from chemical precipitation. Geothermal turbines must be overhauled much more frequently than conventional ones.

6.2.3.2 Hot Dry Rock as an Energy Source

The above section deals with fluid thermal reservoirs which have heat stored in the rock, water and steam. The fluids act as transporters of heat providing the reservoir or aquifer permeability is sufficiently high. If permeability is too low and/or there is no water in the thermal reservoir then techniques have to be found to extract the heat from the hot dry rock [e.g. 6.49]. This heat can only be brought to the surface with the help of a fluid. For this purpose heat exchange surfaces must be made at depth which make circulation of an appropriate fluid possible. Because of poor thermal conductivity of rock, large heat surfaces are necessary. Such surfaces can be enhanced by hydraulic fracturing of the potential reservoir rock. This is a technique which has been employed for many years in the petroleum industry to increase permeability of the source rock. Within a sealed-off portion of a drill hole, fluid pressure is increased until the rock fractures. The resultant fracture, a few millimeters wide, is kept open by circulating material such as quartz sand in the pressurized fluid. Unfortunately the course and orientation of the fracture system which is to be used as a heat exchange surface is often unpredictable. A second drill hole is necessary to extract water which is heated as it migrates from one drillhole to the other through the fractures (Fig. 6.16). A second possibility for creating large heat exchange surfaces is to set off an explosion within the drill hole. Within the immediate surrounding rock it is pulverized and at a distance numerous cracks are formed. This process increases the porosity and permeability [e.g. 6.2], but the latter to a lesser extent.

The first method of hydraulic fracturing is being developed and tested at a number of research projects in various countries. They are trying to develop a method of controlling or directing the fractures. In the vicinity of Los Alamos/USA there is an experimental geothermal power station which has produced electricity from hot dry rock [6.49]. The hydro-fractured reservoir has a temperature about $T = 300\,°C$. Injected water vaporizes at depth and the resulting steam either powers the generator directly or after passing through a secondary system.

Heat extraction from rock made permeable through an explosion requires only one borehole. A pipe of smaller diameter than the drill hole is required through which cold water is pumped down the fragmented reservoir region. This water heats up and

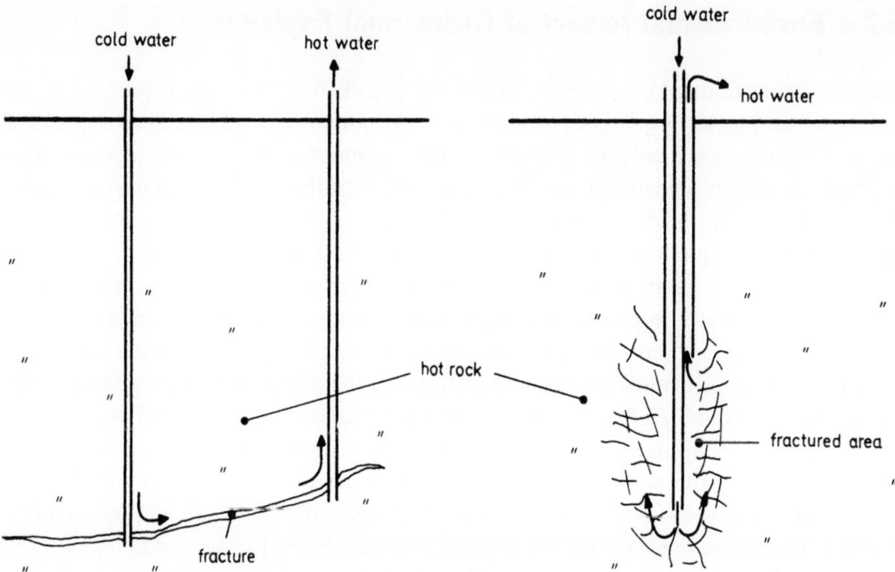

Fig. 6.16. Two possibilities for extracting geothermal heat from dry hot rock

at sufficiently high temperature vaporizes. The steam rises to the surface in the annules between the pipe and wall of the drill hole. A number of connected cracks [e.g. 6.17] are equally suited for this method. Steam can be piped to a turbine directly or through a heat exchanger.

6.2.3.3 Exploitation of Hot Water

Thermal reservoirs with temperatures high enough to produce steam directly or from dry hot rock (after injecting cold water) are relatively rare compared to thermal anomalies of medium temperature. Temperature always increase with increasing depth so that at a certain point sufficiently high temperatures are reached to vaporize water. However, the cost of drilling, which increases disproportionally with depth, is a limiting factor.

If the geothermal energy stored in moderate temperature water is to be converted to electricity, then a secondary system containing a fluid with a low boiling point is necessary. This secondary system containing, for example freon or isobutane together with the turbine, is a single closed unit. Heat extracted from hot water using a heat exchange system evaporates the organic fluid and the vapor then drives the turbine [6.2]. One experimental plant of this type exists in Paratuke/USSR with an output of 0.75 MW.

At this time not much effort is being put into technical improvements of such two-stage systems. Rather, emphasis is placed on drilling sufficiently deep holes so that steam rather than hot water is obtained. The main reason for this emphasis is that the electric power that one can produce from moderate temperature systems is quite limited.

6.2.4 Environmental Impact of Geothermal Exploitation

Increased environmental awareness in the last decade has had a significant influence on the development of geothermal industries. Although environmental concern is of general importance it affects geothermal development in some specific ways. For example, if certain precautions are not observed, then the surface discharge of gases, fluids and heat can lead to harmful biological effects. Similarly surface damage can result from the depletion of the subsurface water reservoir.

Hot water and steam drawn from depth may have strong environmental impacts [6.2, 6.15]. Generally, economically exploitable thermal reservoirs occur in volcanically active regions where the hydrothermal systems not only contain meteoric water but also fossil and juvenile waters exist. Although the meteoric water contains originally almost no solute, it dissolves potentially harmful mineral components (salts and quartz) from the country rock, especially when it is heated. As it rises it transports these minerals to the surface. Fossil water can contain large amounts of salt. For example, groundwater of the Rhinegraben commonly contains a salt content of a few percent. Juvenile water, originating as a differentiation product as a magma evolves, often contains poisonous substances such as hydrogen sulfide, arsenic, mercury, etc.

Harmful materials in water or steam must be separated before thermal water is injected back into a groundwater aquifer or steam into the atmosphere. Often a more practicable method is to pump thermal waste water back into the thermal aquifer itself. However, this must be done at a suitable distance from the extraction boreholes in order to avoid a cooling of the thermal source. This reinjection method can be used for thermal water as well as condensed steam. In regions having a dry hot climate and thermal water with a high mineral content, it may be possible to evaporate the waste water and commercially utilize the salts obtained.

Not only thermal water but also steam may contain noxious chemicals such as hydrogen sulfide, boron, arsenic, ammonia gas, mercury and fluorine (see Table 6.3), all of which can harm the environment in various ways.

CO_2 and H_2S are present in steam in the largest amounts. In 1975 in The Geysers/ USA geothermal field, 28 tons of H_2S were produced daily with the steam [6.44]. With increased production of power emission of H_2S increase. However, H_2S can be technically separated and converted to sulfur. CO_2 emissions in geothermal power stations generally are much below those of conventional power plants using fossil fuel. Yet attention should be given to lower the CO_2 emission climatic changes are possible with higher CO_2 content in the air [6.2]. A very high CO_2 emission characterizes Monte Amiata/Italy.

Table 6.3. Amount of foreign substances [ppm] in steam condensation from three geothermal anomalies of Japan as in [6.28]

Location	HBO_2	NH_4	As	Hg	F
Matsukawa	2	17	0,03	0,06	1,1
Onikobe	1	2	–	0,3	0,6
Otake	0,3	0,5	0,01	0,02	0,2

Steam can also contain tiny particles of SiO_2, which if not filtered out can silicify the surrounding vegetation, as happened in one case in Wairakei/New Zealand [6.2].

Because of low thermal efficiency hot water leaves the turbine at a fairly high temperature. If the water is not further utilized then it must be cooled to below 40 °C in cooling towers before draining it into the surface water. This, however, is not a problem specific to geothermal power plants but also for conventional as well as nuclear plants.

While hot water can be extracted without any real noticeable sound, boreholes producing steam operate with considerable noise caused by the emission of the rapidly expanding steam. Sound absorbers are used to bring the noise level to tolerable level.

Exploitation of water from a confined reservoir can lead to ground subsidence [6.50]. This causes environmental damage as is well known in mining. However it can be prevented by re-injecting the water.

Each geothermal area discovered so far has had specific environmental characteristics which have to be dealt with individually. But generally the environmental problems occurring in geothermal exploitation are ones that can be solved.

7 Appendix

7.1 Error Function

$$\mathrm{erf}(x) = \frac{2}{\sqrt{\pi}} \int\limits_0^x \exp(-t^2)\, dt$$

x	erf(x)	x	erf(x)	x	erf(x)
0.000	0.0000	0.005	0.0056	0.010	0.0113
0.015	0.0169	0.020	0.0226	0.025	0.0282
0.030	0.0338	0.035	0.0395	0.040	0.0451
0.045	0.0507	0.050	0.0564	0.055	0.0620
0.060	0.0676	0.065	0.0732	0.070	0.0789
0.075	0.0845	0.080	0.0901	0.085	0.0957
0.090	0.1013	0.095	0.1069	0.100	0.1125
0.105	0.1180	0.110	0.1236	0.115	0.1292
0.120	0.1348	0.125	0.1403	0.130	0.1459
0.135	0.1514	0.140	0.1569	0.145	0.1625
0.150	0.1680	0.155	0.1735	0.160	0.1790
0.165	0.1845	0.170	0.1900	0.175	0.1955
0.180	0.2009	0.185	0.2064	0.190	0.2118
0.195	0.2173	0.200	0.2227	0.205	0.2281
0.210	0.2335	0.215	0.2389	0.220	0.2443
0.225	0.2497	0.230	0.2550	0.235	0.2604
0.240	0.2657	0.245	0.2710	0.250	0.2763
0.255	0.2816	0.260	0.2869	0.265	0.2922
0.270	0.2974	0.275	0.3027	0.280	0.3079
0.285	0.3131	0.290	0.3183	0.295	0.3235
0.300	0.3286	0.305	0.3338	0.310	0.3389
0.315	0.3440	0.320	0.3491	0.325	0.3542
0.330	0.3593	0.335	0.3643	0.340	0.3694
0.345	0.3744	0.350	0.3794	0.355	0.3844
0.360	0.3893	0.365	0.3943	0.370	0.3992
0.375	0.4041	0.380	0.4090	0.385	0.4139
0.390	0.4187	0.395	0.4236	0.400	0.4284
0.405	0.4332	0.410	0.4380	0.415	0.4427
0.420	0.4475	0.425	0.4522	0.430	0.4569
0.435	0.4616	0.440	0.4662	0.445	0.4709
0.450	0.4755	0.455	0.4801	0.460	0.4847
0.465	0.4892	0.470	0.4937	0.475	0.4983

x	erf(x)	x	erf(x)	x	erf(x)
0.480	0.5027	0.485	0.5072	0.490	0.5117
0.495	0.5161	0.500	0.5205	0.505	0.5249
0.510	0.5292	0.515	0.5336	0.520	0.5379
0.525	0.5422	0.530	0.5465	0.535	0.5507
0.540	0.5549	0.545	0.5591	0.550	0.5633
0.555	0.5675	0.560	0.5716	0.565	0.5757
0.570	0.5798	0.575	0.5839	0.580	0.5879
0.585	0.5919	0.590	0.5959	0.595	0.5999
0.600	0.6039	0.605	0.6078	0.610	0.6117
0.615	0.6156	0.620	0.6194	0.625	0.6232
0.630	0.6270	0.635	0.6308	0.640	0.6346
0.645	0.6383	0.650	0.6420	0.655	0.6457
0.660	0.6494	0.665	0.6530	0.670	0.6566
0.675	0.6602	0.680	0.6638	0.685	0.6673
0.690	0.6708	0.695	0.6743	0.700	0.6778
0.705	0.6812	0.710	0.6847	0.715	0.6881
0.720	0.6914	0.725	0.6948	0.730	0.6981
0.735	0.7014	0.740	0.7047	0.745	0.7079
0.750	0.7112	0.755	0.7144	0.760	0.7175
0.765	0.7207	0.770	0.7238	0.775	0.7269
0.780	0.7300	0.785	0.7331	0.790	0.7361
0.795	0.7391	0.800	0.7421	0.805	0.7451
0.810	0.7480	0.815	0.7509	0.820	0.7538
0.825	0.7567	0.830	0.7595	0.835	0.7623
0.840	0.7651	0.845	0.7679	0.850	0.7707
0.855	0.7734	0.860	0.7761	0.865	0.7788
0.870	0.7814	0.875	0.7841	0.880	0.7867
0.885	0.7893	0.890	0.7918	0.895	0.7944
0.900	0.7969	0.905	0.7994	0.910	0.8019
0.915	0.8043	0.920	0.8068	0.925	0.8092
0.930	0.8116	0.935	0.8139	0.940	0.8163
0.945	0.8186	0.950	0.8209	0.955	0.8232
0.960	0.8254	0.965	0.8277	0.970	0.8299
0.975	0.8231	0.980	0.8342	0.985	0.8364
0.990	0.8385	0.995	0.8406	1.000	0.8427
1.005	0.8448	1.010	0.8468	1.015	0.8488
1.020	0.8508	1.025	0.8528	1.030	0.8548
1.035	0.8567	1.040	0.8586	1.045	0.8606
1.050	0.8624	1.055	0.8643	1.060	0.8661
1.065	0.8680	1.070	0.8698	1.075	0.8716
1.080	0.8733	1.085	0.8751	1.090	0.8768
1.095	0.8785	1.100	0.8802	1.105	0.8819
1.110	0.8835	1.115	0.8852	1.120	0.8868
1.125	0.8884	1.130	0.8900	1.135	0.8915
1.140	0.8931	1.145	0.8946	1.150	0.8961
1.155	0.8976	1.160	0.8991	1.165	0.9006
1.170	0.9020	1.175	0.9034	1.180	0.9048
1.185	0.9062	1.190	0.9076	1.195	0.9090
1.200	0.9103	1.205	0.9116	1.210	0.9130
1.215	0.9143	1.220	0.9155	1.225	0.9168
1.230	0.9181	1.235	0.9193	1.240	0.9205
1.245	0.9217	1.250	0.9229	1.255	0.9241
1.260	0.9252	1.265	0.9264	1.270	0.9275
1.275	0.9286	1.280	0.9297	1.285	0.9308

x	erf(x)	x	erf(x)	x	erf(x)
1.290	0.9319	1.295	0.9330	1.300	0.9340
1.305	0.9350	1.310	0.9361	1.315	0.9371
1.320	0.9381	1.325	0.9390	1.330	0.9400
1.335	0.9410	1.340	0.9419	1.345	0.9428
1.350	0.9438	1.355	0.9447	1.360	0.9456
1.365	0.9464	1.370	0.9473	1.375	0.9482
1.380	0.9490	1.385	0.9499	1.390	0.9507
1.395	0.9515	1.400	0.9523	1.405	0.9531
1.410	0.9539	1.415	0.9546	1.420	0.9554
1.425	0.9561	1.430	0.9569	1.435	0.9576
1.440	0.9583	1.445	0.9590	1.450	0.9597
1.455	0.9604	1.460	0.9611	1.465	0.9617
1.470	0.9624	1.475	0.9630	1.480	0.9637
1.485	0.9643	1.490	0.9649	1.495	0.9655
1.500	0.9661	1.505	0.9667	1.510	0.9673
1.515	0.9678	1.520	0.9684	1.525	0.9690
1.530	0.9695	1.535	0.9701	1.540	0.9706
1.545	0.9711	1.550	0.9716	1.555	0.9721
1.560	0.9726	1.565	0.9731	1.570	0.9736
1.575	0.9741	1.580	0.9745	1.585	0.9750
1.590	0.9755	1.595	0.9759	1.600	0.9763
1.605	0.9768	1.610	0.9772	1.615	0.9776
1.620	0.9780	1.625	0.9784	1.630	0.9788
1.635	0.9792	1.640	0.9796	1.645	0.9800
1.650	0.9804	1.655	0.9807	1.660	0.9811
1.665	0.9815	1.670	0.9818	1.675	0.9822
1.680	0.9825	1.685	0.9828	1.690	0.9832
1.695	0.9835	1.700	0.9838	1.705	0.9841
1.710	0.9844	1.715	0.9847	1.720	0.9850
1.725	0.9853	1.730	0.9856	1.735	0.9859
1.740	0.9861	1.745	0.9864	1.750	0.9867
1.755	0.9869	1.760	0.9872	1.765	0.9874
1.770	0.9877	1.775	0.9879	1.780	0.9882
1.785	0.9884	1.790	0.9886	1.795	0.9889
1.800	0.9891	1.805	0.9893	1.810	0.9895
1.815	0.9897	1.820	0.9899	1.825	0.9901
1.830	0.9903	1.835	0.9905	1.840	0.9907
1.845	0.9909	1.850	0.9911	1.855	0.9913
1.860	0.9915	1.865	0.9916	1.870	0.9918
1.875	0.9920	1.880	0.9922	1.885	0.9923
1.890	0.9925	1.895	0.9926	1.900	0.9928
1.905	0.9929	1.910	0.9931	1.915	0.9932
1.920	0.9934	1.925	0.9935	1.930	0.9937
1.935	0.9938	1.940	0.9939	1.945	0.9941
1.950	0.9942	1.955	0.9943	1.960	0.9944
1.965	0.9945	1.970	0.9947	1.975	0.9948
1.980	0.9949	1.985	0.9950	1.990	0.9951
1.995	0.9952	2.000	0.9953	2.005	0.9954
2.010	0.9955	2.015	0.9956	2.020	0.9957
2.025	0.9958	2.030	0.9959	2.035	0.9960
2.040	0.9961	2.045	0.9962	2.050	0.9963
2.055	0.9963	2.060	0.9964	2.065	0.9965
2.070	0.9966	2.075	0.9967	2.080	0.9967
2.085	0.9968	2.090	0.9969	2.095	0.9970

x	erf(x)	x	erf(x)	x	erf(x)
2.100	0.9970	2.105	0.9971	2.110	0.9972
2.115	0.9972	2.120	0.9973	2.125	0.9973
2.130	0.9974	2.135	0.9975	2.140	0.9975
2.145	0.9976	2.150	0.9976	2.155	0.9977
2.160	0.9977	2.165	0.9978	2.170	0.9979
2.175	0.9979	2.180	0.9980	2.185	0.9980
2.190	0.9980	2.195	0.9981	2.200	0.9981
2.205	0.9982	2.210	0.9982	2.215	0.9983
2.220	0.9983	2.225	0.9983	2.230	0.9984
2.235	0.9984	2.240	0.9985	2.245	0.9985
2.250	0.9985	2.255	0.9986	2.260	0.9986
2.265	0.9986	2.270	0.9987	2.275	0.9987
2.280	0.9987	2.285	0.9988	2.290	0.9988
2.295	0.9988	2.300	0.9989	2.305	0.9989
2.310	0.9989	2.315	0.9989	2.320	0.9990
2.325	0.9990	2.330	0.9990	2.335	0.9990
2.340	0.9991	2.345	0.9991	2.350	0.9991
2.355	0.9991	2.360	0.9992	2.365	0.9992
2.370	0.9992	2.375	0.9992	2.380	0.9992
2.385	0.9993	2.390	0.9993	2.395	0.9993
2.400	0.9993	2.405	0.9993	2.410	0.9993
2.415	0.9994	2.420	0.9994	2.425	0.9994
2.430	0.9994	2.435	0.9994	2.440	0.9994
2.445	0.9995	2.450	0.9995	2.355	0.9995
2.460	0.9995	2.465	0.9995	2.470	0.9995
2.475	0.9995	2.480	0.9995	2.485	0.9996
2.490	0.9996	2.495	0.9996		

7.2 Answer to the Problems

1.1 1.) $\Delta T = -A/K$

    2.) $\quad A = A_0(1 - z/2H)$

$$\frac{d^2 T}{dz^2} = A_0(z/2H - 1)/K$$

$$\frac{dT}{dz} = A_0 z^2/(4HK) + Q_0/K - A_0(H + z)/K$$

$$T = T_0 + Q_0 z/K - (Hz - z^2/2 + z^3/12H)\, A_0/K$$

1.2 1.) $Q_{ME} = Q_0 - A_0 H(1 - \exp(-z/H))$

$$= Q_0 - 30.9$$

    2.) $Q_{ML} = Q_0 - 31.5$

$$\Delta Q = Q_{ME} - Q_{ML} = -0.6\ \text{mW/m}^2$$

1.3 1.) $T_{ME} = T_0 + (Q_0 - A_0 H (1 - \exp(-z/H))) \, z/K$
$$+ A_0 H^2 (1 - \exp(-z/H))/K$$

2.) $T_{ML} = T_0 + (Q_0 - A_0 H) \, z/K + (z^2/2 - z^3/12H) \, A_0/K$

$$\Delta T_M = T_{ML} - T_{ME} = (- Hz \exp(-z/H) + z^2/2 - z^3/12H$$
$$- H^2 (1 - \exp(-z/H)) \, A_0/K$$
$$= 22 \,°K$$

1.4 $\quad T_{MC} = T_0 + Q_0 z/K - A_0 H z/2K$

$$\Delta T = T_{MC} - T_{ME} = -\left(\left(1 - \exp\left(-\frac{z}{H} \right) \right) (z - H) - \frac{z}{2} \right)$$
$$= 74 \,°K$$

2.1 Model A:

$$\Delta K = \frac{3 \cdot 0.25 \left(1 - \dfrac{0.65}{6.1} \right) \cdot 6.1}{2 + 0.25 + 0.65/6.1} = 1.74 \text{ W/m} \,°K$$

Model B:

$$\Delta K = 3.04 \text{ W/m} \,°K$$

2.2 Minimum value:

$$\frac{0.92}{2.5} + \frac{0.08}{5.1} = \frac{1}{2.61} \text{ m} \,°K/W$$

Maximum value:

$$0.92 \cdot 2.5 + 0.08 \cdot 5.1 = 2.71 \text{ W/m} \,°K$$

The difference is:

$$0.11 \leqq \Delta K \leqq 0.21 \text{ W/m} \,°K$$

2.3 Model:

	depth [km]		heat generation [μW/m^3]	$\int A \, dz \left[\dfrac{mW}{m^2} \right]$
Norway	0 17	(upper crust)	1.7 0.7	20.4
	17 36	(lower crust)	0.5 0.13	6.0
Denmark	0 6	(sed. cover)	1.3	7.8
	6 13	(basement)	1.9	13.3
	13 30	(lower crust)	0.6 0.13	6.2

2.4 For the model, see 2.3

$$Q = \int A_0 \exp(-az)\,dz$$

		$\int A\,dz\,[mW/m^2]$
Norway:	upper crust	19.2
	lower crust	5.2
Denmark:	sed. cover	7.8
	basement	13.3
	lower crust	5.2

3.1 From Fig. 3.2 follows $a = 1.65$

$$x = a\sqrt{\frac{2Kt}{c\varrho}} = 1.65\sqrt{\frac{2 \cdot 2 \cdot 10 \cdot 3.15 \cdot 10^7}{1 \cdot 2.5 \cdot 10^6}} = 37\ m$$

3.2 From Fig. 3.4 find $a = \sqrt{2\kappa t}/h \approx 2$

at $z/h - 2$

$$t = \frac{2h^2}{\kappa} = 25\ \text{years}$$

3.3

z	$y = \dfrac{T - T_2}{T_1 - T_2}$	$T = T_2 + y(T_1 - T_2)$
D	0.5	650
2D	0.24	380
3D	0.16	310

3.4 From Fig. 3.9 read $a = 0.9$

$$a = \frac{\kappa t}{R^2}; \quad t = \frac{aR^2}{\kappa} = \frac{2.5^2 \cdot 10^6 \cdot 0.9}{32} = 1.8 \cdot 10^5\ \text{years}$$

3.5 From Fig. 3.9 read at $\dfrac{T - T_2}{T_1 - T_2} = \dfrac{50}{650} = 0.08$

$a = 1.0$

$$t = \frac{aR^2}{\kappa} = \frac{10^6}{32} = 3 \cdot 10^4\ \text{years}$$

4.1 Three steps are applied

$$\sum_1^3 (1 - 2/r_i^2)/(1 + 1/r_i^2)^{2.5}\,\frac{T_{B,i}\,\Delta R}{r_i^2} = \left(\frac{dT}{dz}\right)_c$$

$\Delta r = 0.5\ km$

$r_i\,[km]$	$T_{B,i}\,[°C]$
0.25	17.08
0.75	10.25
1.25	3.42

124

$$\left(\frac{\mathrm{d}T}{\mathrm{d}z}\right)_c = -\frac{31}{1192}\cdot 136.6 - \frac{2.56}{12.86}\cdot 9.11 - \frac{0.28}{3.44}\cdot 1.09$$

$$= -5.45\,^{\circ}\mathrm{C/km}$$

$$\left(\frac{\mathrm{d}T}{\mathrm{d}z}\right)_0 = 25 + 5.5 = 30.5\,^{\circ}\mathrm{C/km}$$

4.2 $$\frac{T - T_1}{T_2} = \frac{200}{6000} + \frac{2}{\pi}\left\{\mathrm{arctg}\,\frac{\sin\dfrac{\pi}{30}}{\mathrm{sh}\dfrac{\pi}{3}} - \mathrm{arctg}\,\frac{\sin\dfrac{\pi}{30}}{\exp\dfrac{\pi}{3} + \cos\dfrac{\pi}{30}}\right\}$$

$$= 0.033 + 0.034$$

$$T = T_1 + 0.067\cdot 80 = T_1 + 5.4$$

4.3 $$\Delta T_i = Q\cdot h_i\left(\frac{1}{K_i} - \frac{1}{K_s}\right)$$

$$\Delta T_1 = 80\cdot 10^{-3}\cdot 200\left(\frac{1}{3.2} - \frac{1}{3.2}\right) = 0$$

$$\Delta T_2 = 1.4$$

$$\Delta T_3 = 2.3$$

$$\Delta T_4 = 2.8$$

$$\Delta T = \sum \Delta T_i = 6.5\,^{\circ}\mathrm{C}\ \text{at the basis of the sedimentary rocks}$$

5.1 $\quad F_1(T) = 2.18 - 0.16 = 2.02 \quad$ with $\quad$ 300 ppm Na

$$T_1 = 113.5\,^{\circ}\mathrm{C}$$

$$F_2(T) = 3.00 - 0.02 = 2.98 \quad \text{with}\ \ 2000\ \text{ppm Na}$$

$$T_2 = 42.5\,^{\circ}\mathrm{C}$$

$$\Delta T = 71\,^{\circ}\mathrm{C}$$

5.2 $$R_m^2 = f(\mathrm{d}T/\mathrm{d}z)\int_0^{t_1} z(t)\,\mathrm{d}t$$

$$0.4^2 = f(\mathrm{d}T/\mathrm{d}z)\int_0^{20}\frac{1.5}{20}\,t\,\mathrm{d}t$$

$$0.4^2 = f(\mathrm{d}T/\mathrm{d}z)\cdot 15$$

$$I = \frac{15}{0.16} = 93.75\ \text{km Ma}$$

$$\frac{\mathrm{d}T}{\mathrm{d}z} = 32.4\,^{\circ}\mathrm{C/km}$$

5.3 $$R_m^2 = f(\mathrm{d}T/\mathrm{d}z)\left\{\int_{t_1}^{t_2}\left(z_0 + \frac{\Delta z}{\Delta t}\right)\mathrm{d}t + c\right\}$$

$$1.)\ \ 1.35^2 = f(\mathrm{d}T/\mathrm{d}z)\cdot c$$

2.) $\quad 1.8^2 = f(dT/dz) \left\{ \int_0^{2.4} \left(2.14 + \frac{0.55}{2.4} t \right) dt + c \right\}$

$\qquad\qquad = f(dT/dz) \{5.80 + c\}$

From 1.) and 2.) follows $c = 7.46$

$$I = \frac{7.46 + 5.80}{1.8^2} = 4.09 \text{ km Ma}$$

$$\frac{dT}{dz} = 78 \,°C/km$$

5.4 $\quad T_\infty^1 - 93.5 = (T_\infty^1 - 92) \exp(-c^1(268 - 200))$

$\qquad T_\infty^1 - 95 \quad = (T_\infty^1 - 92) \exp(-c^1(404 - 200))$

$\qquad T_\infty^2 - 97 \quad = (T_\infty^2 - 96) \exp(-c^2(656 - 500))$

$\qquad T_\infty^2 - 98 \quad = (T_\infty^2 - 96) \exp(-c^2(968 - 500))$

$\qquad T_\infty^1 = 95.9 \,°C; \quad c^1 = 7.14 \cdot 10^{-3} \min^{-1}$

$\qquad T_\infty^2 = 98.6 \,°C; \quad c^2 = 3.11 \cdot 10^{-3} \min^{-1}$

For $c \to 0$ follows $T_\infty = 100.7 \,°C$

8 References

References to the Introduction

[1] Aepinus, F. A.: De distributione calor. per tellum, St. Petersburg 1761.
[2] Bischof, G.: Die Wärmelehre des Innern unseres Erdkörpers, Leipzig 1837.
[3] Buch, L. von: Einige Bemerkungen über Quellen-Temperatur, Ann. Phys. Chem. 88 (= 12 Neue Folge), 403–418, 1828.
[4] Buffon, Gr. von: Epochen der Natur, aus dem Französischen Les époches de la nature (Paris 1780), Leipzig 1782.
[5] Cassini de Thury, J. D.: Sur la température des souterrains de l'observatoire royal, Mémoires présentés par divers savants à l'académie Française, p. 511, 328–329, 1786.
[6] Descartes, R.: Opera mathematica et philosophica, tom. I, principia philosophiae, Amsterdam 1692.
[7] Fourier, M.: Théorie analytique de la chaleur, Paris 1822, dtsche Ausg. von B. Weinstein, Analytische Theorie der Wärme, Berlin 1884.
[8] Hire, Ph. de la: in [5] Cassini de Thury.
[9] Holmes, A.: Radioactivity and the Earth's thermal history, Geol. Mag. 52, 60–71 und 102–115, 1915.
[10] Hopkins, W.: Researches in Geology, Philos. Trans. Roy. Soc. Lond. 129, 381–423, 1839; ibd. 130, 193–208, 1840; ibd. 132, 43–55, 1842.
[11] Humboldt, A. von: Kosmos, Bd. 1, 340 ff., Stuttgart und Tübingen 1845.
[12] Hunt, St.: The chemistry of the primeval Earth, Geol. Mag. 5, 49–59, 1868.
[13] Ingersoll, L. R. & Zobel, O. I.: Mathematical theory of heat conduction, Boston 1913.
[14] Jeffreys, H.: On the Earth's thermal history and some related geological phenomena, Gerl. Beitr. Geophys. 18, 1–29, 1927.
[15] Kircher, A.: Mundus subterranus, Amsterdam 1665.
[16] Klöden, K. F.: Über die Zunahme der Temperatur nach dem Innern der Erde, Jhrb. Min., Geogn., Geol. u. Petref. 2, 385–390, 1831.
[17] Kupffer, A. T.: Über die mittlere Temperatur der Luft und des Bodens auf einigen Punkten des östlichen Rußlands, Ann. Phys. Chem. 91 (= 15 Neue Folge), 159–192, 1829.
[18] Leibnitz, G. W. von: Protogaea sive de prima facie telluris et antiquissimae historie vestigiis, Göttingen 1749.
[19] Liebenow, C.: Notiz über die Radiummmenge der Erde, Phys. Ztschr. 5, 625–626, 1904.
[20] Lyell, Ch.: Principles of geology, 3 Bde., London 1830–1833.
[21] Newton, I.: Philosophiae naturalis principia mathematica (1687), dtsche Ausg. von J. Ph. Wolfers, mathematische Prinzipien der Naturlehre, Berlin 1872.
[22] Parrot, G. F.: Grundriß der Physik der Erde und Geologie, Riga und Leipzig 1815.
[23] Rive, A. de la: in: Poisson, Von den Ursachen der Temperatur des Erdballs, Ann. Phys. Chem. 115 (= 39 Neue Folge), 66–100, 1836.
[24] Strutt, R. J.: On the radioactive minerals, Proc. Roy. Soc. Lond., Ser. A, 76, 88–101, 1905.
[25] Thiene, H.: Temperatur und Zustand des Erdinnern, Jena 1907.

For Advanced Studies to: 1. Physical Basis of Heat Transfer

Carslaw, H .S. & Jaeger, J. C.: Conduction of heat in solids, 2nd. ed., Oxford 1952.
Tautz, H.: Wärmeleitung und Temperaturausgleich, Weinheim/Bergstr. 1971.

References to: 2. Thermal Properties of Common Rocks

[2.1] Balling, N. P.: Geothermal models of the crust and the uppermost mantle of the Fenno-scandian shield in south Norway and the Danish embayment, J. Geophys. 42, 237–256, 1976.

[2.2] Birch, F. & Clark, H.: The thermal conductivity of rocks and its dependence upon temperature and composition, Am. J. Sci. 238, 529–558, 613–635, 1940.

[2.3] Bridgman, P.W.: The physics of high pressure, London (Bell & Sons) 1952.

[2.4] Buntebarth, G.: Geophysikalische Untersuchungen über die Verteilung von Uran, Thorium und Kalium in der Erdkruste sowie deren Anwendung auf Temperaturberechnungen für verschiedene Krustentypen, Diss. TU Clausthal, Clausthal-Zellerfeld 1975.

[2.5] Buntebarth, G.: Methoden zur Abschätzung der Wärmeflußdichte aus dem oberen Mantel, Geol. Rdschau 65, 809–819, 1976.

[2.6] Buntebarth, G. & Rybach, L.: Linear relationships between petrophysical properties and mineralogical constitution – preliminary results, Tectonophys. 75, 41–46, 1981.

[2.7] England, P.C.: Some thermal considerations of the Alpine metamorphism – past, present and future, Tectonophys. 46, 21–40, 1978.

[2.8] Fielitz, K.: Untersuchungen zur Temperaturabhängigkeit von Kompressions- und Scherwellengeschwindigkeiten in Gesteinen unter erhöhtem Druck, Diss. TU Clausthal, Clausthal-Zellerfeld 1971.

[2.9] Fukao, Y., Mizutani, H. & Uyeda, S.: Optical absorption spectra at high temperatures and radiative thermal conductivity of olivines, Phys. Earth Planet. Int. 1, 57–62, 1968.

[2.10] Grubbe, K.; Hänel, R. & Zoth, G.: Determination of the vertical components of thermal conductivity by line source methods, in: Hänel, R. & Gupta (eds.), Results of the first workshop on standards in geothermics, 49–56, Stuttgart (Schweizerbart) 1983.

[2.11] Holmes, A.: Radioactivity and the Earth's thermal history, Geol. Mag. 52, 60–71 und 102–112, 1915.

[2.12] Horai, K. & Simmons, G.: Thermal conductivity of rock forming minerals, Earth Planet. Sci. Lett. 6, 359–368, 1969.

[2.13] Hurtig, E. & Brugger, H.: Wärmeleitfähigkeitsmessungen unter einaxialem Druck, Tectonophys. 10, 67–77, 1970.

[2.14] Kanamori, H., Fuji, N. & Mizutani, H.: Thermal diffusivity measurement of rock-forming minerals from 300 to 1100 °K, J. Geophys. Res. 73, 595–605, 1968.

[2.15] Kappelmeyer, O. & Hänel, R.: Geothermics with special reference to applicaton, Berlin-Stuttgart (Gebrüder Borntraeger), 1974.

[2.16] Kawada, K.: Studies of the thermal state of the Earth. The 15th paper: Variation of thermal conductivity of rocks, Bull. Earthquake Res. Inst. Tokyo 42, 631–647, 1964.

[2.17] Physikhütte (Hrsg.: Hütte Gesellschaft f. Technische Information), Bd. II, 29. Aufl., S. 392 (Ernst & Sohn), Berlin-München-Düsseldorf, 1971.

[2.18] Kobayashi, Y.: Anisotropy of thermal diffusivity in olivine, pyroxene and dunite, J. Phys. Earth 22, 359–373, 1974.

[2.19] Labhart, T. P. & Rybach, L.: Granite und Uranvererzungen in den Schweizer Alpen, Geolog. Rundschau 63, 135–147, 1974.

[2.20] Nafe, J. E. & Drake, C. L.: in: M. Talwani, G. H. Sutton, J. L. Worzel: A crustal section across the Puerto Rico Trench, J. Geophys. Res. 64, 1548, 1959.

[2.21] Rybach, L.: Wärmeproduktionsbestimmungen an Gesteinen der Schweizer Alpen, Beiträge zur Geologie der Schweiz, Geotechn. Serie, Lieferung 51, 1973.

[2.22] Rybach, L.: Radioactive heat production in rocks and its relation to other petrophysical parameters, Pageoph. 114, 309–317, 1976.

[2.23] Rybach, L. & Buntebarth, G.: Heat generating radioelement in granitic magmas, J. Volcan. Geotherm. Resources 10, 395–404, 1981.

[2.24] Sass, J. H.: The thermal conductivity of fifteen feldspar specimens, J. Geophys. Res. 70, 4064–4065, 1965.

[2.25] Schatz, J. F. & Simmons, G.: Thermal conductivity of earth materials at high temperatures, Journ. of Geophys. Res. 77, 6966–6983, 1972.

[2.26] Schloessin, H. H. & Dvořák, Z.: Anisotropic Lattice thermal conductivity in Enstatite as a function of pressure and temperature, Geophys. J.R. Astr. Soc. 27, 499–516, 1972.

128

[2.27] Schmucker, U.: Geophysical aspects of structure and composition of the earth. in: K. H. Wedepohl (ed.), Handbook of Geochemistry, Vol. 1, 134–226 (Springer), Berlin-Heidelberg-New York, 1969.

[2.28] Seibold, U. & Gutzeit, W.: Untersuchungen der Druckabhängigkeit der Wärmeleitfähigkeit einiger Gesteine, Gerl. Beitr. Geophys. 83, 498–504, 1974.

[2.29] Staudacher, W.: Die Temperatur-Leitfähigkeit von natürlichem Olivin bei hohen Drucken und Temperaturen, Ztschr. f. Geophysik 39, 979–988, 1973.

[2.30] Van der Molen, I.: The shift of the $\alpha - \beta$ transition temperature of quartz associated with the thermal expansion of granite at high pressure, Tectonophys. 73, 323–342, 1981.

[2.31] Wakita, H., Nagasawa, H., Uyeda, S. & Kuno, H.: Uranium, thorium and potassium contents of possible mantle materials, Geochem. J. 1, 183–198, 1967.

[2.32] Walsh, J. B. & Decker, E. R.: Effect of pressure and saturating fluid on the thermal conductivity of compact rock, J. Geophys. Res. 71, 3053–3061, 1966.

[2.33] Wenk, H.-R. & Wenk, E.: Physical constants of Alpine rocks (density, porosity, specific heat, thermal diffusivity and conductivity), Schweiz. Mineral. und Petrogr. Mitt. 49, 343–357, 1969.

For Advanced Studies to: 3. Analytical Treatment of Cooling in the Crust

Carslaw, H .S. & Jaeger, I. C.: Conduction of heat in solids, 2nd. ed., London (Oxford Univ. Press) 1959.

Lovering, T. S.: Theory of heat conduction applied to geological problems, Bull. Geol. Soc. Am. 46, 69–94, 1935.

Mundry, E.: Über die Abkühlung magmatischer Körper, Geol. Jb. 85, 755–766, 1968.

Tautz, H.: Wärmeleitung und Temperaturausgleich, Weinheim/Bergstr. (Verlag Chemie) 1971.

References to: 4. Thermal State of the Earth's Interior

[4.1] Anderson, D. L.: Composition of the mantle and core, Ann. Rev. Earth Planet. Sci. 5, 179–202, 1977.

[4.2] Baer, A. J.: Geotherms, evolution of the lithosphere and plate tectonics, Tectonophys. 72, 203–227, 1981.

[4.3] Balke, K. D.: Geothermische und hydrogeologische Untersuchungen in der südlichen Niederrheinischen Bucht, Geol. Jb., C, Heft 5, Hannover 1973.

[4.4] Balling, N. P.: Geothermal models of the crust and the uppermost mantle of the Fennoscandian shield in south Norway and the Danish embayment, J. Geophys. 42, 237–256, 1976.

[4.5] Birch, F.: Flow of heat in the Front Range, Colorado, Bull. Geol. Soc. Am. 61, 567–630, 1950.

[4.6] Bodmer, Ph., England, P. C., Kissling, E. & Rybach, L.: On the correction of subsurface temperature measurements for the effects of topographic relief, part II: Application to temperature measurements in the central Alps, in: V. Čermák & L. Rybach (Hrsg.), Terrestrial heat flow in Europe, 78–87, (Springer), Berlin-Heidelberg-New York 1979.

[4.7] Borchert, H.: Zur Petrologie der Lithosphere in ihrer Beziehung zu geophysikalischen Diskontinuitäten, auch der Gesamterde, Gerl. Beitr. Geophys. 76, 257–277, 1967.

[4.8] Boschi, E.: Melting of iron, Geophys. J. Roy. Astron. Soc. 38, 327–334, 1974.

[4.9] Bullard, E. C.: The disturbance of the temperature gradient in the earth's crust by inequalities of height, Month. Not. Roy. Astron. Soc., Geophys. suppl. 4, 360–362, 1940.

[4.10] Buntebarth, G.: Geophysikalische Untersuchungen über die Verteilung von Uran, Thorium und Kalium in der Erdkruste sowie deren Anwendung auf Temperaturberechnungen für verschiedene Krustentypen, Diss. TU Clausthal, Clausthal-Zellerfeld 1975.

[4.11] Buntebarth, G.: Methoden zur Abschätzung der Wärmeflußdichte aus dem oberen Mantel, Geol. Rdschau 65, 809–819, 1976.

[4.12] Buntebarth, G.: The degree of metamorphism of organic matter in sedimentary rocks as a paleogeothermometer, applied to the Upper Rhinegraben, in: L. Rybach & L. Stegena (Hrsg.), Geothermics and Geothermal Energy, 83–91 (Birkhäuser), Basel 1978.

[4.13] Buntebarth, G. & Schopper, J. R.: Heat flow caused by water migration along faults in dependence on petrophysical parameters, Proc. Int. Congr. Thermal Waters, Geotherm. Energy and Vulcan. Mediterr. Area, Oct. 5–10, 1976; Vol. II, 41–49, Athen 1976.

[4.14] Buntebarth, G. & Teichmüller, R.: Zur Ermittlung der Paläotemperaturen im Dach des Bramscher Intrusivs aufgrund von Inkohlungsdaten, Fortschr. Geol. Rhld. u. Westf. 27, 171–182, 1979.

[4.15] Čermák, V.: Heat flow map of Europe, in: V. Čermák & L. Rybach (Hrsg.), Terrestrial heat flow in Europe, 3–40, (Springer), Berlin-Heidelberg-New York 1979.

[4.16] Chapman, D. S., Pollack, H. N. & Čermák, V.: Global heat flow with special reference to the region of Europe, in: V. Čermák & L. Rybach (Hrsg.), Terrestrial heat flow in Europe, 41–48, (Springer), Berlin-Heidelberg-New York 1979.

[4.17] Crough, S. T. & Thompson, G. A.: Thermal model of continental lithosphere, J. Geophys. Res. 81, 4857–4862, 1976.

[4.18] Davis, E. E. & Lister, C. R. B.: Fundamentals of ridge topography, Earth Planet. Sci. Lett. 21, 405–413, 1974.

[4.19] Davis, E. E. & Lister, C. R. B.: Heat flow measured over the Juan de Fuca ridge: evidence for widespread hydrothermal circulation in a highly heat transportive crust, J. Geophys. Res. 82, 4845–4860, 1977.

[4.20] Davies, G. F.: Review of oceanic and global heat flow estimates, Rev. Geoph. Space Phys. 18, 718–722, 1980.

[4.21] Dietz, R. S.: Continent and ocean basin evolution by spreading of the sea floor, Nature 190, 854–857, 1961.

[4.22] Dorf, E.: The use of fossil plants in paleoclimatic interpretation, in: A. E. M. Nairn (ed.), 13–31, (Interscience), London-New York-Sydney 1964.

[4.23] Eaton, J. P. & Murata, K. J.: How volcanos grow, Science 132, 925–938, 1960.

[4.24] England, P. C.: On the correction of subsurface temperature measurements for the effects of topographic relief, part I: Corrections in terrains with high relief, in: V. Čermák & L. Rybach (Hrsg.), Terrestrial heat flow in Europe, 74–77, (Springer), Berlin-Heidelberg-New York 1979.

[4.25] Fowler, A. C.: On the thermal state of the earth's mantle, J. Geophys. 53, 42–51, 1983.

[4.26] Frenzel, B.: Die Klimaschwankungen des Eiszeitalters, (Vieweg), Braunschweig 1967.

[4.27] Giesel, W. & Holz, A.: Das anomale geothermische Feld in Salzstöcken – Quantitative Deutung an einem Beispiel, Kali u. Steinsalz 5, 272–274, 1970.

[4.28] Gilvarry, J. J.: Temperatures in the Earth's interior, J. Atmosph. Terrestr. Phys. 10, 84, 1957.

[4.29] Graham, E. K. & Dobrzykowski, D.: Temperatures in the mantle as inferred from simple compositional models, Am. Mineral. 61, 549–559, 1976.

[4.30] Hänel, R.: Untersuchungen zur Bestimmung der terrestrischen Wärmestromdichte in Binnenseen, Diss. TU Clausthal, Clausthal-Zellerfeld 1968.

[4.31] Hänel, R.: Eine neue Methode zur Bestimmung der terrestrischen Wärmestromdichte in Binnenseen, Ztschr. Geophys. 36, 725–742, 1970.

[4.32] Hänel, R.: A critical review of heat flow measurements in sea and lake bottom sediments, in: V. Čermák & L. Rybach (Hrsg.), Terrestrial heat flow in Europe, 49–73, (Springer), Berlin-Heidelberg-New York 1979.

[4.33] Hänel, R. & Gupta, M. (eds.): Results of the first workshop on standards in geothermics, Stuttgart (Schweizerbart) 1983.

[4.34] Hess, H. H.: History of ocean basins, in: Engel, A. E. J., H. L. James & Leonard, B. F. (eds.), Petrologic studies, a volume in honor of A. F. Buddington, 599–620, (Geol. Soc. Am.) Boulder 1962.

[4.35] Higgins, G. & Kennedy, G. C.: The adiabatic gradient and the melting point gradient in the core of the Earth, J. Geophys. Res. 76, 1870–1878, 1971.

[4.36] Holmes, A.: Radioactivity and Earth movements, Trans. Geol. Soc. Glasgow 18, III, 559–606, 1931.

[4.37] Holmes, A.: The machinery of continental drift: the search for a mechanism, in: Principles of physical geology, 505–509 (Nelson) London 1944.

[4.38] Honda, S. & Yeda, S.: Thermal process beneath subduction zones, abstract in: IUGG interdisciplinary symposia, Vol. 1, 23, IUGG XVIII General Assembly, Hamburg 1983.

[4.39] Horváth, F., Bodri, L. & Ottlik, P.: Geothermics of Hungary and the tectonophysics of the Pannonian basin "red spot", in: V. Čermák & L. Rybach (Hrsg.), Terrestrial heat flow in Europe, 206–217, (Springer), Berlin-Heidelberg-New York 1979.

[4.40] Hurtig, E. & Čermák, V.: Mapping of the heat flow pattern in Europe, Rev. Roum. Géol. Géophys. et Géogr. Géophysique 22, 73–82, Bucarest 1978.

[4.41] Hurtig, E. & Oelsner, Chr.: Heat flow, temperature distribution and geothermal models in Europe: some tectonic implications, Tectonophys. 41, 147–156, 1977.

[4.42] Jacobs, J. A.: The Earth's core, (Acad. Press) London-New York-San Francisco 1975.

[4.43] Jaupart, C., Sclater, J. G. & Simmons, G.: Heat flow studies: constraints on the distribution of uranium, thorium and potassium in the continental crust, Earth Planet. Sci. Lett. 52, 328–344, 1981.

[4.44] Jeffreys, H.: The disturbance of the temperature gradient in the Earth's crust by inequalities of height, Mont. Not. Roy. Astron. Soc. Geophys. suppl. 4, 309–312, 1940.

[4.45] Kappelmeyer, O. & Hänel, R.: Geothermics with special reference to application, Berlin-Stuttgart (Gebrüder Borntraeger) 1974.

[4.46] Kertz, W.: Einführung in die Geophysik I, (Bibliographisches Institut) Mannheim 1969.

[4.47] Lachenbruch, A. H.: Rapid estimation of the topographic disturbance to superficial thermal gradients, Rev. Geophys. 6, 365–400, 1968.

[4.48] Lees, C. H.: On the shapes of the isogeotherms under mountain ranges in radioactive districts, Proc. Roy. Soc. Ser. A 83, 339–346, 1910.

[4.49] Le Pichon, X.: Sea-floor spreading and continental drift, J. Geophys. Res. 73, 3661–3705, 1968.

[4.50] Lindquist, G.: Heat flow density measurements in the bottom sediment of some lakes in north Sweden, Doctoral Thesis, (1983:30D), Luleå Univ., Luleå/Sweden 1983.

[4.51] Lister, C. R. B.: Estimators for heat flow and deep rock properties based on boundary layer theory, Tectonophys. 41, 157–171, 1977.

[4.52] Morgan, W. J.: Convection plumes in the lower mantle, Nature 230, 42–43, 1971.

[4.53] Möller, F.: Einführung in die Meteorologie, Bd. 2: Physik der Atmosphäre (Bibliographisches Institut) Mannheim-Wien-Zürich 1973.

[4.54] Mundry, E.: Berechnung des gestörten geothermischen Feldes mit Hilfe eines Relaxationsverfahrens, Z. Geophys. 32, 157–162, 1966.

[4.55] Parker, R. L. & Oldenburg, D. W.: Thermal model of ocean ridges, Nature Phys. Sci. 242, 137–139, 1973.

[4.56] Parson, B. & Sclater, J. G.: An analysis of the variation of ocean floor bathymetry and heat flow with age, J. Geophys. Res. 82, 803–827, 1977.

[4.57] Pilger, A., Rösler, A. & Schwan, W.: Zeitlich-tektonische Zusammenhänge bei der Plattentektonik, Clausthaler Geol. Abh. 17, (E. Pilger), Clausthal-Zellerfeld 1974.

[4.58] Pilger, A. & Rösler, A.: Afar between continental and oceanic rifting, (Schweizerbart) Stuttgart 1976.

[4.59] Pollack, H. N. & Chapman, D. S.: Mantle heat flow, Earth Planet. Sci. Lett. 34, 174–184, 1977.

[4.60] Polyak, B. G. & Smirnov, Ya. B.: Heat flow on continents, Doklady of Acad. Sci. USSR, Earth Sci. Sect. 168 (engl. transl.) 26–29, 1966.

[4.61] Reynolds, R. T. & Summers, A. L.: Calculations on the composition of the terrestrial planets, J. Geophys. Res. 74, 2494–2511, 1969.

[4.62] Ringwood, A. E.: Phase transformations and the constitution of the mantle, Phys. Earth Planet. Int. 3, 109–155, 1970.

[4.63] Ringwood, A. E. & Major, A.: The system $Mg_2SiO_4-Fe_2SiO_4$ at high pressures and temperatures, Phys. Earth Planet. Int. 3, 89–108, 1970.

[4.64] Roy, R. F., Blackwell, D. D. & Birch, F.: Heat generation of plutonic rocks and continental heat flow provinces, Earth Planet. Sci. Lett. 5, 1–12, 1968.

[4.65] Rybach, L. & Muffler, L. J. P.: Geothermal systems – principles and case histories, Chichester-New York-Brisbane-Toronto (J. Wiley & Sons) 1981.

[4.66] Savin, S. M.: The history of the Earth's surface temperature during the past 100 million years, Ann. Rev. Earth Planet. Sci. 5, 319–355, 1977.

[4.67] Schönenberg, R.: Einführung in die Geologie Europas (Rombach) Freiburg 1971.

[4.68] Schroth, G. J.: A new probe for the in situ determination of the heat flow density in shallow cased boreholes, in: Hänel, R. & M. Gupta (eds.), Results of the 1st-workshop on standards in geothermics, 5–16, Stuttgart (Schweizerbart) 1983.

[4.69] Schubert, G., Froidevaux, C. & Yuen, D. A.: Oceanic lithosphere and asthenosphere: thermal and mechanical structure, J. Geophys. Res. 81, 3525–3540, 1976.

[4.70] Schwarzbach, M.: Das Klima der Vorzeit – Eine Einführung in die Paläoklimatologie, 3. Aufl., (Enke) Stuttgart 1974.

[4.71] Sclater, J. G. & Crowe, J.: On the reliability of oceanic heat flow averages, J. Geophys. Res. 81, 2997–3006, 1976.

[4.72] Sclater, J. G.; Crowe, J. & Anderson, R. N.: On the reliability of oceanic heat flow averages, J. Geophys. Res. 81, 2997–3006, 1976.

[4.73] Sclater, J. G. & Francheteau, J.: The implications of terrestrial heat flow observations on current tectonic and geochemical models of the crust and upper mantle of the Earth, Geophys. J. Roy. Astron. Soc. 20, 509–542, 1970.

[4.74] Sclater, J. G., Anderson, R. N. & Bell, M. L.: Elevation of ridges and evolution of the central eastern Pacific, J. Geoph. Res. 76, 7888–7915, 1971.

[4.75] Sclater, J. G., Jaupart, C. & Galson, D.: The heat flow through oceanic and continental crust and the heat loss of the earth, Rev. Geophys. Space Phys. 18, 269–311, 1980.

[4.76] Smith, G. D.: Numerische Lösung von partiellen Differentialgleichungen, (Vieweg) Braunschweig 1970.

[4.77] Solomon, S. C.: Geophysical constraints on radial and lateral temperature variations in the upper mantle, Am. Mineral. 61, 788–803, 1976.

[4.78] Stacey, F. D.: Physical properties of the Earth's core, Geophys. Surv. 1, 99–119, 1972.

[4.79] Stacey, F. D.: A thermal model of the Earth, Phys. Earth Planet. Int. 15, 341–348, 1977.

[4.80] Stegena, L.: Migration und Geothermik im Ungarischen Becken, Vortr. IV. Int. Wiss. Conf. Chem. u. Phys. Probleme d. Erkundung u. Förderung von Erdöl und Erdgas, I, 115–119, Prag 1966.

[4.81] Tautz, H.: Wärmeleitung und Temperaturausgleich, Weinheim/Bergstr. (Verlag Chemie) 1971.

[4.82] Teichmüller, M. & Teichmüller, R.: Zur geothermischen Geschichte des Oberrhein-Grabens. Zusammenfassung und Auswertung eines Symposiums, Fortschr. Geol. Rhld. u. Westf. 27, 109–120, 1979.

[4.83] Tozer, D. C.: The electrical properties of the Earth's interior, in: L. H. Ahrens et al. (Hrsg.), Physics and Chemistry of the Earth, Vol. 3, 414–436, (Pergamon Press) New York 1959.

[4.84] von Herzen, R. P. & Maxwell, A.: The measurement of thermal conductivity of deep sea sediments by a needle probe method, J. Geophys. Res. 64, 1557–1563, 1959.

[4.85] Werner, D.: Probleme der Geothermik am Beispiel des Rheingrabens, Diss. Univ. Karlsruhe, Karlsruhe 1975.

[4.86] Werner, D. & Kley, W.: Problems of heat storage in aquifers, J. Hydrol. 34, 35–43, 1977.

[4.87] Werner, D. & Parini, M.: The geothermal anomaly of Landau/Pfalz: An attempt of interpretation, J. Geophys. 48, 28–33, 1980.

[4.88] Wyllie, P. J.: Crustal anatexis: an experimental review, Tectonophys. 43, 41–71, 1977.

[4.89] Yefimov, A. V., Kutasov, I. M. & Shipitsina, L. I.: Zone of influence of ground water sources on the temperature field of bottom deposits, Izv., Earth Phys. (engl. transl.), 90–95, 1975.

References to: 5. Methods for Determining Temperatures

[5.1] Ádám, A.: Results of deep electromagnetic investigations, in: A. Ádám (Hrsg.), Geoelectrical and geothermic studies, 547–560, (Akad. Kiadó) Budapest 1976.

[5.2] Albright, J.: A new and more accurate method for the direct measurement of earth temperature gradient in deep boreholes, Proc. 2nd U.N. Intern. Symp., Development and Use of Geothermal Resources, Vol. 2, 847–851, Washington 1976.

132

[5.3]   Angenheister, G. & Soffel, H.: Gesteinsmagnetismus und Paläomagnetismus, Studienhefte zur Physik des Erdkörpers 1, (Borntraeger) Berlin-Stuttgart 1972.

[5.4]   Archie, G. E.: The electrical resistivity log as an aid in determinating some reservoir characteristics, Trans. AIME, Petrol. Br. 146, 54–62, 1942.

[5.5]   Banno, S. & Matsui, Y.: Eclogite types and partition of Mg, Fe, and Mn between clinopyroxene and garnet, Proc. Japan Acad. Tokyo 41, 716–721, 1965.

[5.6]   Bethke, P. M. & Barton jr, P. B.: Distribution of some minor elements between coexisting sulfide minerals, Economic Geol. 66, 140–163, 1971.

[5.7]   Borchert, M.: Ozeane Salzlagerstätten, (Borntraeger) Berlin 1959.

[5.8]   Bosum, W., Hahn, A., Kind, E. G. & Pucher, P.: Geomagnetic anomalies in geothermal areas – Rhine Graben and Urach area –, Seminar on Geothermal Energy, Vol. I, 277–295, (Kommission der Europäischen Gemeinschaften, EUR 5920), Brüssel 1977.

[5.9]   Boyd, F. R.: A pyroxene geotherm, Geochim. Cosmochim. Acta 27, 2533–2546, 1973.

[5.10] Braitsch, O. & Herrmann, A. G.: Zur Geochemie des Broms in salinaren Sedimenten, Teil II: Die Bildungstemperaturen primärer Sylvin- und Carnallit-Gesteine, Geochim. Cosmochim. Acta. 28, 1081–1109, 1964.

[5.11] Bullard, E. C.: The time necessary for a borehole to attain temperature equilibrium, Geophys. Suppl. to the Month. Not. R.A.S. London 5, 127–130, 1947.

[5.12] Buntebarth, G.: Über die Größe der thermisch bedingten Bouguer-Anomalie in den Alpen, Ztschr. f. Geophys. 39, 109–114, 1973.

[5.13] Buntebarth, G.: The degree of metamorphism of organic matter in sedimentary rocks as a paleogeothermometer, applied to the Upper Rhine Graben, in: L. Rybach & L. Stegena (Hrsg.), Geothermics and Geothermal Energy, 83–91, (Birkhäuser) Basel 1978.

[5.14] Buntebarth, G.: Eine empirische Methode zur Berechnung von paläogeothermischen Gradienten aus dem Inkohlungsgrad organischer Einlagerungen in Sedimentgesteinen mit Anwendung auf den mittleren Oberrheingraben, Fortschr. Geol. Rhld. u. Westf. 27, 97–108, 1979.

[5.15] Buntebarth, G.: Geothermal history estimated from the coalification of organic matter, Tectonophys. 83, 101–108, 1982.

[5.16] Buntebarth, G. & Teichmüller, M.: Ancient heat flow density estimated from the coalification of organic matter in the borehole Urach 3 (SW-Germany) in: Hänel, R. (ed.), The Urach geothermal project, 89–95, Stuttgart (Schweizerbart) 1982.

[5.17] Buntebarth, G., Koppe, I. & Teichmüller, M.: Paleogeothermics in the Ruhr basin, in: V. Čermák & R. Hänel (eds.), Geothermics and geothermal energy, 45–55, Stuttgart (Schweizerbart) 1982.

[5.18] Čermák, V.: Heat flow in the Upper Silesian coal basin, Pageoph. 69, 119–130, 1968.

[5.19] Epstein, S., Buchsbaum, R., Lowenstam, H. A. & Urey, H. C.: Revised carbonate-water isotopic temperature scale, Bull. Geol. Soc. Am. 64, 1315–1325, 1953.

[5.20] Fielitz, K.: Elastische Wellengeschwindigkeiten in verschiedenen Gesteinen unter hohem Druck und bei Temperaturen bis 750 °C, Z. f. Geophys. 37, 943–956, 1971.

[5.21] Fournier, R. O.: Application of water geochemistry to geothermal exploration and reservoir engineering, in: Rybach, L. & Muffler, L. J. P.: Geothermal Systems: principles and case histories, 109–143, Chichester-New York-Brisbane-Toronto (John Wiley & Sons) 1981.

[5.22] Fournier, R. O. & Rowe, J. J.: Estimation of underground temperatures from silica content of water from hot springs and wet-steam wells, Am. J. Sci. 264, 685–697, 1966.

[5.23] Fournier, R. O. & Truesdell, A. H.: An empirical Na–K–Ca geothermometer for natural waters, Geochim. Cosmochim. Acta 37, 1255–1276, 1973.

[5.24] Fournier, R. O., White, D. E. & Truesdell, A. H.: Geochemical indicators of subsurface temperatures – part 1, basic assumptions, J. Res. US Geol. Survey 2, 259–262, 1974.

[5.25] Giese, P.: Die Temperaturverteilung in der Erdkruste des Alpenvorlandes und der Alpen, abgeschätzt aus tiefenseismischen Beobachtungen, Schweiz. Min. Petr. Mitt. 50, 597–610, 1970.

[5.26] Goldsmith, J. R. & Newton, R. C.: P–T–X relations in the system $CaCO_3 – MgCO_3$ at high temperatures and pressures, Am. J. Sci. 267-A, 160–190, 1969.

[5.27] Goldstein, N. E. & Paulsson, B.: Interpretation of gravity surveys in Grass and Buena Vista valleys, Nevada, Geothermics 7, 29–50, 1978.

[5.28] Günther, R., Kappelmeyer, O. & Kronberg, P.: Zur Prospektion auf geothermale Anomalien, Erfahrungen einer Modelluntersuchung in Polichnitos, Lesbos (Griechenland), Geol. Rundsch. 66, 10–33, 1977.

[5.29] Hänel, R.: Bericht über die Berechnung und Darstellung von Temperaturen aus geochemischen Thermometern auf dem Gebiet der Bundesrepublik Deutschland, Nieders. Landesamt f. Bodenforschung, Arch. Nr. 78249, Hannover 1977.

[5.30] Hahn, A., Kind, E. G. & Mishra, D. C.: Depth estimation of magnetic sources by means of fourier amplitude spectra, Geophys. Prosp. 24, 287–308, 1976.

[5.31] Harker, R. I. & Tuttle, O. F.: Studies in the system $CaO-MgO-CO_2$, P.2: Limits of solid solution along the binary join, $CaCO_3-MgCO_3$, Am. J. Sci. 235, 274–282, 1955.

[5.32] Hedemann, H.-A.: Beiträge zur Geothermik aus Tiefbohrungen, Freib. Forsch.-hefte, C 238: Methoden und Ergebnisse geothermischer Untersuchungen, 63–77, 1968.

[5.33] Hoefs, J.: Stable isotope geochemistry, (Springer) Berlin-Heidelberg-New York 1973.

[5.34] Hoover, D. B., Long, C. L. & Senterfit, R. M.: Audiomagnetotelluric investigations in geothermal areas, Geophys. 43, 1501–1514, 1978.

[5.35] Irving, A. J.: Geochemical and high pressure experimental studies of garnet pyroxenite and pyroxene granulite xenoliths from the Delegate basaltic pipes, Australia, J. Petrol. 15, 1–40, 1974.

[5.36] Irving, A. J.: On the validity of paleogeotherms determined from xenolith suites in basalts and kimberlites, Am. Mineral. 61, 638–642, 1976.

[5.37] Jin, D. J.: True-temperature determination of geothermal reservoirs, Geoexplor. 15, 1–9, 1977.

[5.38] Kahle, H. G. & Werner, D.: Gravity and temperature anomalies in the wake of drifting continents, Tectonophys. 29, 487–504, 1975.

[5.39] Kappelmeyer, O. & Hänel, R.: Geothermics with special reference to application, (Borntraeger) Berlin-Stuttgart 1974.

[5.40] Karweil, J.: Die Metamorphose der Kohlen vom Standpunkt der physikalischen Chemie, Z. deutsch. geol. Ges. 107, 132–139, 1955.

[5.41] Kolesar, P. T. & Degraff, J. V.: A comparison of the silica and $Na-K-Ca$ geothermometers for thermal springs in Utah, Geothermics 6, 221–226, 1978.

[5.42] Lachenbruch, A. H. & Brewer, H. C.: Dissipation of the temperature effect of drilling a well in arctic Alaska, Bull. Geol. Survey N 1083-C, 73, 1959.

[5.43] Lee, T.-Ch.: On shallow-hole temperature measurements – a test study in the Salton Sea geothermal field, Geophys. 42, 572–583, 1977, 8 fig., 3 tabl.

[5.44] Lopatin, N. W.: Temperature and geologic time as factor in coalification (engl. transl.), Acad. Nauk SSR, Izv., Ser. Geol. 3, 95–106, 1971.

[5.45] MacGregor, I. D.: The system $MgO-Al_2O_3-SiO_2$: Solubility of Al_2O_3 in enstatite for spinel and garnet peridotite compositions, Am. Mineral. 59, 110–119, 1974.

[5.46] McKenzie, D.: The variation of temperature with time and hydrocarbon maturation in sedimentary basins formed by extension, Earth Planet. Sci. Lett. 55, 87–98, 1981.

[5.47] Middleton, M. F.: The subsidence and thermal history of the Bass basin, southeastern Australia, Tectonophys. 87, 383–397, 1982.

[5.48] Middleton, M. F. & Schmidt, P. W.: Paleothermometry of the Sydney basin, J. Geophys. Res. 87, B, 5351–5359, 1982.

[5.49] Mori, T. & Green, D. H.: Subsolidus equilibria between pyroxenes in the $CaO-MgO-SiO_2$ system at high pressures and temperatures, Am. Mineral. 61, 616–625, 1976.

[5.50] Nielson, H.: Sulfur isotopes, in: E. Jäger & J. C. Hunziker (Hrsg.), Lectures in isotope geology, 283–312, (Springer) Berlin-Heidelberg-New York 1979.

[5.51] Ogawa, K.: Gravimetric survey at the southern part of Izu, preliminary survey for geothermal exploration, Bull. Geol. Survey Japan 28, 35–44, 1977.

[5.52] Ohara, M. J. & Yarwood, G.: High pressure-temperature point of a Archaean geotherm, implied magma genesis by crustal anatexis, and consequences for garnet-pyroxene thermometry and barometry, Phil. Trans. R. Soc. London A 288, 441–456, 1978.

[5.53] O'Neil, J. R.: Stable isotope geochemistry of rocks and minerals, in: E. Jäger & J. C. Hunziker (Hrsg.), Lectures in isotope geology, 235–263, (Springer) Berlin-Heidelberg-New York 1979.

134

[5.54] Powell, R.: The thermodynamics of pyroxene geotherms, Phil. Trans. R. Soc. London A 288, 457–469, 1978.
[5.55] Puhan, D.: Metamorphic temperature determined by means of the dolomite-calcite solves geothermometer – examples from the central Damara orogen (South West Africa), Contrib. Mineral. Petrol. 58, 23–28, 1976.
[5.56] Puhan, D.: Metamorphism of siliceous dolomites of the central and southern part of the Damara orogen, in: H. Martin & F. W. Eder (eds.), Intracontinental fold belts, 767–784, Berlin-Heidelberg-New York-Tokyo (Springer) 1983.
[5.57] Quist, A. S. & Marshall, W. I.: Electrical conductances of aqueous sodium chloride solutions from 0 to 800 °C and at pressures to 4000 bars, J. Phys. Chem. 72, 684–706, 1968.
[5.58] Râheim, A. & Green, D. H.: Experimental determination of the temperature and pressure dependence of the Fe – Mg partition coefficient for coexisting garnet and clinopyroxene, Contrib. Mineral. Petrol. 48, 179–203, 1974.
[5.59] Rink, M. & Schopper, J. R.: Interface conductivity of saturated porous media and its relation to structure, Proc. RILEM-IUPAC Intern. Symp. "Pore Structure and properties of materials", final report, Vol. 2, C/311–C/320, Prag 1973.
[5.60] Sabins, F. F. jr.: Remote sensing – principles and interpretation, (Freeman) San Francisco 1978.
[5.61] Schmucker, U.: Conductivity anomalies, with special reference to the Andes, in: Runcorn, S. K. (Hrsg.), The application of modern physics to the earth and planetary interiors, 125–138, (Wiley-Interscience) London-New York-Sydney-Toronto 1969.
[5.62] Shuey, R. T., Schellinger, D. K., Tripp, A. C. & Alley, L. B.: Curie depth determination from areomagnetic spectra, Geophys. J. Roy. Astr. Soc. 50, 75–101, 1977.
[5.63] Sommer, J.: Der Einfluß der Bildung fluider Phasen in natürlichen Gesteinen auf das Ausbreitungsverhalten elastischer Wellen, Diss. TU Clausthal, Clausthal-Zellerfeld 1979.
[5.64] Swanberg, Ch. A. & Morgan, P.: The linear relation between temperatures based on the silica content of ground water and regional heat flow: A new heat flow map of the United States, Pageoph. 117, 227–241, 1978.
[5.65] Taylor, H. P. jr.: The application of oxygen and hydrogen isotope studies to problems of hydrothermal alteration and ore deposition, Econ. Geol. 69, 843–883, 1974.
[5.66] Teichmüller, M.: Bestimmung des Inkohlungsgrades von kohligen Einschlüssen in Sedimenten des Oberrheingrabens – ein Hilfsmittel bei der Klärung geothermischer Fragen, in: J. H. Illies & St. Müller (Hrsg.), Graben Problems, 124–142, (Schweizerbart) Stuttgart 1970.
[5.67] Teichmüller, M. & Teichmüller, R.: Zur geothermischen Geschichte des Oberrhein-Grabens. Zusammenfassung und Auswertung eines Symposiums, Fortschr. Geol. Rhld. u. Westf. 27, 109–120, 1979.
[5.68] Teichmüller, M. & Teichmüller, R.: The significance of coalification studies to geology – a review, Bull. Centres Explor.-Prod. Elf-Aquitaine 5, 491–534, 1981.
[5.69] Tissot, B.: Premières données sur les mécanismes et la cinétique de la formation du pétrole dans les sédiments; simulation d'un schéma réactionnel sur ordinateur, Rev. Inst. Franç. Petrole 24, 470–501, 1969.
[5.70] Truesdell, A. H.: Summary of section III, Geochemical techniques in exploration, Proc. 2nd U.N. Symp. on the Development and Use of Geothermal Resources, Vol. I, 53–79, Washington 1976.
[5.71] Usdowski, H.-E.: Fraktionierung der Spurenelemente bei der Kristallisation, (Springer) Berlin-Heidelberg-New York 1975.
[5.72] Volarovich, M. P. & Parkhomenko, E. I.: Electrical properties of rocks at high temperatures and pressures, in: A. Ádám (Hrsg.), Geoelectrical and geothermal studies, 321–369, Budapest 1976.
[5.73] Wohlenberg, J.: Geophysikalische Untersuchungen in der Forschungsbohrung Urach, Progr. Energieforsch. u. Energietechnol. 1977–1980, Statusreport 1978 – Geotechnik und Lagerstätten, Bd. 1, 87–100, (Projektleitung Energieforsch., KFA Jülich), Jülich 1978.
[5.74] Wohlenberg, J. & Hänel, R.: Kompilation von Temperatur-Daten für den Temperatur-Atlas der Bundesrepublik Deutschland, Programm Energieforsch. u. Energietechnol. 1977–1980, Statusreport 1978, Geotechnik u. Lagerstätten, Bd. 1, 1–12, (Projektleitung Energieforschung, KFA Jülich), Jülich 1978.

References to: 6. Geothermal Heat as an Energy Source

[6.1] Allen, G. W. & McCluer, H. K.: Abatement of hydrogen sulphide emissions from the Geysers geothermal power plant, Proc. 2nd U.N. Int. Symp., Development and Use of Geothermal Resources, San Francisco 1975, Vol. 2, 1313–1316, Washington 1976.

[6.2] Armstead, H. C. H.: Geothermal Energy, (Spon) London 1978.

[6.3] Axtmann, R. C.: Chemical aspects of the environmental impact of geothermal power, Proc. 2nd U.N. Int. Symp., Development and Use of Geothermal Resources, San Francisco 1975, Vol. 2, 1323 ff., Washington 1976.

[6.4] Balling, N. & Saxov, S.: Low enthalpy geothermal energy resources in Denmark, Pageoph. 117, 205–212, 1978.

[6.5] Barker, C. E. & Elders, W. A.: Vitrinite reflectance geothermometry and apparent heating duration in the Cerro Prieto geothermal field, Geothermics 10, 207–223, 1981.

[6.6] Boldizsár, T.: Non-electric use of geothermal energy in Hungary, Acta Geodaet., Geophys. et Montanist. Acad. Sci. Hung. Tom. 14, 289–297, 1979.

[6.7] Brauchle, A. & Groh, W.: Zur Geschichte der Physiotherapie, (Haug), 4. Aufl., Heidelberg 1971.

[6.8] Buntebarth, G.: The degree of metamorphism of organic matter in sedimentary rocks as a paleogeothermomcter, applied to the Upper Rhine Graben, in: L. Rybach & L. Stegena (Hrsg.), Geothermics and Geothermal Energy, 83–91, (Birkhäuser) Basel 1978.

[6.9] Buntebarth, G. & Schopper, J. R.: Heat flow caused by water migration along faults in dependence on petrophysical parameters, Proc. Int. Congr. Thermal Waters, Geotherm. Energy and Vulcan. Mediterr. Area, Oct. 5–10, 1976; Vol. II, 41–49, Athen 1976.

[6.10] Buntebarth, G., Grebe, H., Teichmüller, M. & Teichmüller, R.: Inkohlungsuntersuchungen in der Forschungsbohrung Urach 3 und ihre geothermische Interpretation, Fortschr. Geol. Rhld. u. Westf. 27, 183–199, 1979.

[6.11] Buntebarth, G. & Teichmüller, R.: Zur Ermittlung der Paläotemperaturen im Dach des Bramscher Intrusivs aufgrund von Inkohlungsdaten, Fortschr. Geol. Rhld. u. Westf. 27, 171–182, 1979.

[6.12] Creutzburg, H.: Untersuchungen über den Wärmestrom der Erde in Westdeutschland, Kali u. Steinsalz 4, 73–108, 1964.

[6.13] Crittenden, M. D., Jr.: Environmental aspects of geothermal development, in: Rybach, L. & Muffler, L. J. P. (eds.), Geothermal systems: principles and case histories, 199–217, Chichester-New York-Brisbane-Toronto (John Wiley & Sons) 1981.

[6.14] Corwin, R. F. & Hoover, D. B.: The self-potential method in geothermal exploration, Geophys. 44, 226–245, 1979.

[6.15] Coulbois, P. & Hérault, J.-P.: Conditions for the competitive use of geothermal energy in home heating, Proc. 2nd U.N. Int. Symp., Development and Use of Geothermal Resources, San Francisco 1975, Vol. 3, 2104 ff., Washington 1976.

[6.16] Eriksson, K. G., Ahlbom, K., Landström, O., Larson, S. Å., Lind, G. & Malmquist, D.: Investigation for geothermal energy in Sweden, Pageoph. 117, 196–204, 1978.

[6.17] Ernst, P. L.: Frac-Studien in der erweiterten Forschungsbohrung Urach, Programm Energieforsch. u. Energietechnol. 1977–1980, Statusreport 1978 – Geotechnik u. Lagerstätten, Bd. 1, 101–109, (Projektleitung Energieforschung, KFA Jülich), Jülich 1978.

[6.18] Fridleifsson, I. B.: Applied volcanology in geothermal exploration in Iceland, Pageoph. 117, 242–252, 1978.

[6.19] Gringarten, A. C.: Reservoir lifetime and heat recovery factor in geothermal aquifers used for urban heating, Pageoph. 117, 297–308, 1978.

[6.20] Gringarten, A. C. & Witherspoon, P. A.: Extraction of heat from multiple-fractured dry hot rock, Geothermics 2, 119–122, 1973.

[6.21] Gudmundsson, J. S.: Low-temperature geothermal use in Iceland, Geothermics 11, 59–68, 1982.

[6.22] Hot Spring Institute of Kanagawa-Prefecture, Map of Bouguer-anomaly of the Kanagawa-Prefecture. 1:100,000, Hakone 1975.

[6.23] Iriyama, J. & Oki, Y.: Thermal structure and energy of the Hakone volcano, Japan, Pageoph. 117, 331–337, 1978.

136

[6.24] Ishii, Y.: Passive and active seismic prospecting in geothermal area, Butsuri-Tankô (Geophys. Explor.) 29, 14–31, 1976.

[6.25] Kappelmeyer, O.: Implications of heat flow studies for geothermal energy prospects, in: V. Čermák u. L. Rybach (Hrsg.), Terrestrial heat flow in Europe, 126–135, (Springer) Berlin-Heidelberg-New York 1979.

[6.26] Kappelmeyer, O. & Hänel, R.: Geothermics with special reference to application, (Borntraeger) Berlin-Stuttgart 1974.

[6.27] Kertz, W.: Kann Erdwärme unseren Energiebedarf decken? Umschau 74, 661–666, 1974.

[6.28] Koga, A.: Geochemistry of geothermal system prospecting, Butsuri-Tankô (Geophys. Explor.) 29, 72–82, 1976.

[6.29] Kruger, P., Stoker, A. & Umana, A.: Radon in geothermal reservoir engineering, Geothermics 5, 13–19, 1977.

[6.30] Lamethe, D. & Laurent, G.: Investigation of the optimal use of geothermal waters for the heating of several types of dwelling in various European climates, Seminar on Geothermal Energy, Vol. 2, 559–570, (Kommission der Europ. Gemeinschaften, EUR 5920), Brüssel 1977.

[6.31] Laughlin, A. W.: The geothermal system of the Jemez Mountains, New Mexico (USA) and its exploration, in: Rybach, L. & Muffler, L. J. P. (eds.), Geothermal systems: principles and case histories, 295–320, Chichester-New York-Brisbane-Toronto (John Wiley & Sons) 1981.

[6.32] Lejeune, J. M.: Opération Creil: Lecons à tirer et premiers bilans, 2éme Coll. Franco-Allemand sur les Recherches geothermiques dans le Fosse Rhenan Sup., Strasbourg 1979, 43–44, (B.R.G.M.), Strasbourg 1980.

[6.33] Majer, E. L. & McEvilly, T. V.: Seismological investigations at the Geysers geothermal field, Geophys. 44, 246–269, 1979.

[6.34] Meidav, T., Sanyal, S. & Facca, G.: An update of world geothermal energy development, Geotherm. Energy Mag. 5, 30–34, 1977.

[6.35] Militzer, H., Schön, J., Stötzner, U. & Stoll, R.: Angewandte Geophysik im Ingenieur- und Bergbau, (VEB Deutscher Verlag f. Grundstoffindustrie) Leipzig 1978.

[6.36] Muffler, L. J. P.: Geothermal resource assessment, in: Rybach, L. & Muffler, L. J. P. (eds.), Geothermal systems: principles and case histories, 181–198, Chichester-New York-Brisbane-Toronto (John Wiley & Sons) 1981.

[6.37] Nakamura, H. & Sumi, K.: Exploration and development at Takinone, Japan, in: Rybach, L. & Muffler, L. J. P. (eds.), Geothermal systems: principles and case histories, 247–272, Chichester-New York-Brisbane-Toronto (John Wiley & Sons) 1981.

[6.38] Nohl, G.: 1892–1972 Deutscher Bäderverband (DBV), (Boldt) Bonn 1972.

[6.39] Nourbehecht, B.: Irreversible thermodynamic effects in inhomogeneous media and their application in certain geoelectric problems, Ph.D. thesis, Mass. Inst. Technol., Cambridge/Mass. 1963.

[6.40] Oelsner, Chr.: Anwendung der Infrarotoberflächengeometrie zur Erkundung von Hohlräumen, Freib. Forsch.-Heft C 341, 155–178, Leipzig (VEB Deut. Verl. f. Grundstoffindustrie) 1979.

[6.41] Oki, Y. & Hirano, T.: Hydrothermal system and seismic activity of Hakone volcano, Proc. US-Lapan cooperative seminar "The Utilization of Volcanic Energy", 153–166, Hilo/Hawaii 1974.

[6.42] Onodera, S.: An estimation of potentials for the Hatchobaru geothermal area, northern Kyushu, Japan, Proc. US-Japan cooperative seminar "The Utilization of Volcanic Energy", 75–105, Hilo/Hawaii 1974.

[6.43] Ottlik, P., Gálfi, J., Horvath, F., Korim, K. & Stegena, L.: The low enthalpy geothermal resource of the Pannomian basin, Hungary, in: Rybach, L. & Muffler, L. J. P. (eds.), Geothermal systems: principles and case histories, 221–245, Chichester-New York-Brisbane-Toronto (John Wiley & Sons) 1981.

[6.44] Reed, M. J. & Campbell, G.: Environmental impact of development in the Geysers geothermal filed, USA, Proc. 2nd U.N. Int. Symp., Development and Use of Geothermal Resources, San Francisco 1975, Vol. 2, 1399 ff., Washington 1976.

[6.45] Rybach, L. & Muffler, L. J. P.: Geothermal systems: principles and case histories, Chichester-New York-Brisbane-Toronto (John Wiley & Sons) 1981.

[6.46] Sabins, F. F. jr.: Remote sensing – principles and interpretation, (Freeman) San Francisco 1978.

[6.47] Schaumberg, G.: Geothermisches Pilot-Projekt Bühl, 2ème Coll. Franco-Allemand sur les Recherches Geothermiques dans de Fosse Rhenan Sup., Strasbourg 1979, 39–40, (B.R.G.M.) Straßburg 1980.

[6.48] Sekioka, M. & Yuhara, K.: Heat flux estimation in geothermal areas based on the heat balance of the ground surface, J. Geophys. Res. 79, 2053–2058, 1974.

[6.49] Smith, M. C.: Heat extraction from hot, dry, crustal rocks, Pageoph. 117, 290–296, 1978.

[6.50] Stilwell, W. B., Hall, W. K. & Tawhai, J.: Ground movement in New Zealand geothermal fields, Proc. 2nd U.N. Int. Symp., Development and Use of Geothermal Resources, San Francisco 1975, Vol. 2, 1427 ff., Washington 1976.

[6.51] Suyuma, J., Sumi, K., Baba, K., Takashima, I. & Yuhara, K.: Assessment of geothermal resources of Japan, Proc. United States-Japan Geol. Surveys panel discussion on the assessment of geothermal resources, Tokyo 1975, 63–119, (Geolog. Survey Japan), Tokyo 1976.

[6.52] Teichmüller, M.: Die Diagenese der kohligen Substanzen in den Gesteinen des Tertiärs und Mesozoikums des mittleren Oberrhein-Grabens, Fortschr. Geol. Rhld. u. Westf. 27, 19–49, 1979.

[6.53] Ward, S. H., Parry, W. T., Nash, W. P., Sill, W. R., Cook, K. L., Smith, R. B., Chapman, D. S., Brown, F. H., Whelan, J. A. & Bowman, J. R.: A summary of the geology, geochemistry and geophysics of the Roosevelt Hot Springs thermal area, Utah, Geophys. 43, 1515–1542, 1978.

[6.54] Werner, D. & Parini, M.: The geothermal anomaly of Landau/Pfalz: An attempt of interpretation, J. Geophys. 48, 28–33, 1980.

[6.55] White, D. E. & Guffanti, M.: Geothermal systems and their energy resources, Rev. Geoph. Space Phys. 17, 887–902, 1979.

[6.56] Yuhara, K. & Ushijima, K.: Ground temperature surveys and thermal discharge measurements at Ibusuki and its surrounding geothermal areas, Bull. Geol. Surv. Japan 28, 33–56, 1977.

Subject Index

absorption coefficient 91, 107
activation energy 77, 89
adiabatic temperature 57–59
age of the earth 32
air temperature 25, 37
albedo 33, 99
Alps 20, 37, 39, 50, 87
alunite 94
ammonia 98, 116
amorphous silica 66, 94
amphibol 18, 62, 75
amplitude of temperature
  variation 33–34, 81
andesite 105
anion 18
anisotropic thermal conductivity 5, 10, 13–14
anisotropy 9, 13–14, 30, 38
anorthite 10
anthracite 77
antigorite 71
aplite 26
aquifer 95, 110, 116
arithmetic mean 10, 21
Arrhenius equation 77
arsen 96–97, 116
aseismic front 62
asthenosphere 50, 57–58, 60–63
atmosphere 33
atmospheric window 99
Australia 80

Basalt 21, 47, 63, 74–76, 80
– melting 89
bathing pool 107–108
bituminous coal 9, 15
Bohemian massif 50
boiling point 87, 108, 110, 115
Boltzmann's constant 89
borehole – exploitation 96, 110–111
– for reference 88
– geological data 78, 98
– location 100, 103, 105
– measurement 46, 79–85
– temperature 35, 39, 81–85
– water injection 42

boron 96–97, 116
Bouguer anomaly 86–87, 103
boundary condition 7
Bramsche intrusiv 50, 77
bromine 71–73
brown coal 77

Cadmium 72–73
Cainozoic 50, 52–53
calcite 76–77
calcium 75, 68–70, 92
Caledonian 49
Cambrium 48
carbon 69
carbon dioxide 96, 116
Carboniferous 80
catazone 71
cation 18
cation packing index 18
Cerro Prieto 98
chalcedony 66–67
chondrite 18
christobalite 66–67, 94
chrysotile 70
clay 15
climatic changes 32, 35–36
clinopyroxene 74–75
coal 43–45, 77–78, 92, 98
coal metamorphism 98
coalification 77–79, 98
compressional wave velocity 18–19, 58,
  91, 107
contact zone 42
continent 11, 46, 49, 54, 60, 63
continental heat flow density 46, 52
convection 16, 32, 40, 50, 57, 60, 63
convective heat 39, 46, 100
cooling 22, 24, 26–31
– crust 52–53, 61
– model 22–29
core 16, 58–60
country rock 22–24, 26, 28, 30–31, 43, 45
crack porosity 14
Cretaceous 80
crust 51, 53–54, 56

140

142

mantle diapir 61
mamtle heat flow density 54
matrix conductivity 89
mazeral 78
melting 56, 85, 89
melting temperature 57–59
mercury 96–97, 116
– thermometer 80, 83
Mesozoic 50, 52, 54
mesozone 71
metamorphic rock 10, 16, 70–71, 74
meteoric water 39, 51, 94–95, 98, 109–110
Mexico 98, 112
mica 13
mica shist 12–13, 18
microcrack 12
micro-earthquake 107
mid ocean ridge 46–47, 52, 60–63
mineral water 109
– fraction 10
Miocene 50
model body 26–31
Mohorovicić discontinuity 8, 48, 50, 57,
    60–61
Monte Amiata 112, 116
morphology 32
muscovite 56

Na-K-Ca-thermometer 68–69
New Zealand 112, 117
nickel 59
Nigorikawa 94
North America 63
Norway 21, 55–56
noxious chemicals 116

Ocean 11, 47, 63
– floor 47, 52–53, 62
– temperature 70
oceanic crust 46–47, 52
– heat flow density 46–47, 52–53, 61
oil 14–15
Oligocene 79
olivine 10–11, 13, 17, 57, 70–71, 75
opal 66
ore 17, 71–73, 98
organic matter 65, 77–80, 92, 94, 98
orthopyroxene 13, 75
Otake 97, 104–106
oxygen 36, 57, 69–71

Paleocene 36
paleogeothermal gradient 80, 92
paleo-heat flow density 80
paleothermometer 36
Paleozoic 49
Pannonian basin 39, 50, 109, 111

Parisian basin 111
partial melting 19, 62–63
peat 77
pegmatite 26
penetration depth 33–35
peridotite 9, 18, 76, 91
permeability 40–41, 97, 109, 114
Permian 80
Permo-Carboniferous 35
pH-value 66, 71
phase transformation 12, 57–59
– quartz 91
phonon conductivity 10
plagioclase 10
plate 26, 30
plate tectonics 60–63
Pleistocene 94
Pliocene 50
Poisson's equation 7–8, 37
pore fluid 15, 21, 62, 88, 97, 105
– space 12, 14–16, 89, 98
– water 109
porosity 12, 14–15, 86–87, 105, 107, 114
porous rock 14, 21, 65, 87–88
potassium 8, 16, 21, 47, 51, 53–54, 59–60,
    68–69, 92
powerplant 111–114, 116–117
Precambrium 35, 47, 52
pressure 19, 21, 40–41, 57, 74–77, 88–89,
    105
Proterozoic 54
pyrite 71
pyrometer 80
pyroxene 57, 74–75

Quartz 10, 12–13, 65–68, 91, 116
Quaternary 35–36

Radiation energy 32–33
– temperature 81, 99–100
radiative conductivity 11
radioactive element 16, 51
radiogenic heat 8, 16–21, 50–51, 54, 60–61
radiometer 99
radium 97
radon 96–97
reaction temperature 75, 78, 98
recharge zone 39
rectangular intrusion 29–30
Red Sea 63
reflectivity 77–80, 92, 98
remanent magnetization 85
resistance thermometer 81, 100
Rhinegraben 40, 50, 61, 68, 79–80, 90, 98,
    109, 111, 116
rifting 50
rock matrix 14, 21, 98

144

Terrestrial Heat Flow in Europe

Editors: **V. Čermák, L. Rybach**

1979. 151 figures, 47 tables. 1 twelve-color
map. VIII, 328 pages
ISBN 3-540-09440-7

"... The essence of this book is the map of the
heat flow in the European region, pragmati-
cally defined here to run from Egypt to the
Ural Mountains to Iceland. The map is in
colors overlying a tectonic basemap. The
colors are well chosen ... the editors have performed a giant task, processing 3076 values of
heat flow with the greatest attention to detail
and in a most timely fashion. Their resulting
book will serve as a handbook for the study
of heat flux in other parts of the world. They
have shown the value of cooperative efforts in
geophysics independent of political boundaries. The essence of their work is the syner-
gism. They are to be congratulated; their
work is a prototype for all future studies of
heat flux on a continental scale.".

Geo Journal

"... I recommend the monograph to all those
interested in thermal processes of the crust
and mantle. It represents a significant contribution to our understanding of a geologically
complex area of the globe, and will provide
stimulus for further geothermal research,
both in Europe and elsewhere."

Tectonophysics

Springer-Verlag
Berlin
Heidelberg
New York
Tokyo

J. S. Rinehart

Geysers and Geothermal Energy

1980. 97 figures, 43 tables. XVI, 223 pages
ISBN 3-540-90489-1

This book presents a comprehensive and systematic account of the geological aspects of geysers and related geothermal phenomena. Richly illustrated, it emphasizes their hydrologic and geologic settings and structures, their function and interaction with the environment. Discussions center on the actions occurring within idealized columnar and pool geysers; current theories are evaluated.

Geysers and Geothermal Energy is unique not only in the depth of its coverage of these fascinating geological and geophysical phenomena, but also in its account of their world-wide occurrence. Geysers and geothermal structures are no longer just of local geologic interest; an understanding of them is important for our increasing use and search for geothermal energy.

Springer-Verlag
Berlin
Heidelberg
New York
Tokyo